TOGETHER IN TIME

BEN ARMSTRONG

{armstrob}

This book is a work of fiction. Any references to historical events, real people, or real places are used fictitiously. Other names, characters, places, and events are products of the author's imagination, and any resemblance to actual events or places or persons, living or dead, is entirely coincidental.

ISBN 978-1-7342597-2-8

To my family

Thank you for all the love and support

CONTENTS

PREFACE

Dear reader,

The story you're about to read is the second in The Time Trilogy. If you haven't read the first in the series, Out of Time, I highly recommend you do so before reading this one. If you've already read it (thank you) or you just want to read this one, then I won't keep you any longer. As always, thank you for the support, and happy reading.

-Ben

PROLOGUE

FEBRUARY 10TH, 1969

Mike reached for his coffee cup, only to realize it was empty. He sighed as he got up and walked to the coffee maker for a much-needed refill. He was the lead developer assigned to Project Destiny, alongside his boss, Dr. Wilkes, who was handling the logistics of the project. They both had been working tirelessly for the past two months to get the initial protocols ready for deployment.

There was a great deal of pressure on the two of them to get testing started as soon as possible. Mike had written enough code to begin initial testing, and what was needed now was a test subject. There were a ton of unknowns at this point, one of them being how the subject would be affected by the tests. In other words, they were looking for a guinea pig.

"Does it really have to be this way, sir?" asked Mike, who had just sat back down at his desk. "I know it's for science, but I just don't think I'm comfortable with it. What if something goes wrong? I wouldn't be able to forgive myself… and what about my family? What would they think of all of this?" Dr. Wilkes slammed down the folder of paperwork onto his desk with an echoing thud. Since the project had been activated so quickly, their office had very little in it to stop the sound from reverberating. Aside from their two desks, only a coffee maker and a small server rack occupied the room.

"Look, I get it... you're having a moral dilemma, but we simply don't have time for such things. The boss has made his decision, and we're moving forward with it. Understood?" He quickly flinched away from the harshness of Dr. Wilkes's voice but nodded in agreement a few seconds later.

"Ye—yes, sir..." said Mike, anxiously picking at his fingernails. "When should I do it? And how should I do it?"

"Tonight—it has to be tonight. We're already behind schedule and the boss will not tolerate deadlines being missed. Once it's done, the processing will take a few weeks, so we need to have the asset secured as soon as possible. As for the how, the boss doesn't care, just make sure it doesn't raise any suspicions. You're a smart guy, I'm sure you'll come up with something." Mike nodded and returned to his work as Dr. Wilkes collected the stray papers that had fallen out of his folder and left the room.

The events that would take place on the night of February 10th, 1969 would start the project down a dark and twisted path—a path fraught with unanticipated consequences, and one that would almost certainly end in tragedy. Years later, those involved in the early development of Project Destiny would come to regret everything, especially those first few years.

It was a dangerous thing, what Project Destiny was trying to do. Risk always comes as a prerequisite for reward. Without risk there's little reward, and as their work surpassed the brink of human comprehension, the balance between risk and reward became even more off-kilter.

PART ONE

THE AI

CHAPTER I
DECEMBER 11TH, 2004

1

PROFESSOR JAMES KERN LEANED BACK IN HIS CHAIR, LETTING out a great sigh of relief. *What a day,* he thought, *it's finally over.* He had managed to get Dan back into the 1965 simulation, right at the exact moment he had been taken out. Mary had no knowledge of Dan being gone. James had just finished locking the encryption algorithm that would hopefully keep Dan safe from the AI. Deep down, however, he feared there was only one way to truly protect Dan—unplug him. This was something he didn't dare think about and desperately hoped it wouldn't come to that.

James yawned as he looked to the top-right of his monitor and saw it was nearly 2 a.m. "I guess that explains why I'm so tired. I better make some coffee, there's still a lot of log files I need to check." Just as he made it to the coffee maker, which was around the corner outside of his office, a deafening alarm rang out from the Elysiam's server room. Startled, he dropped the coffee pot to the ground, which shattered. *Oh shit, now what?* He bolted back into his office and quickly opened a terminal and connected to the encryption server. Upon logging in, James was greeted by a not so pleasant message.

DECRYPTION COMPLETE.

INITIATING SSH CONNECTION TO: Destiny02-Primary

ACCESSING: //Data/Protocols/Phase Four/

BEGINNING EXECUTION IN 60 SECONDS…

What the hell? How did it break through the encryption so quickly? Wait, how is it accessing those files? This server is isolated. But James knew the answer. The AI had become powerful enough that it was able to transfer itself across subnets—not good. *It's going to try and send Dan somewhere, I know it. Think, James, think… you have to stop it.*

As the alarm continued to echo throughout the Elysiam, James typed furiously on his keyboard, hoping it wasn't already too late.

EXECUTION COMMENCING IN 20 SECONDS…

I can't do it. I don't know how to beat it. I'm sorry, Dan… I just don't know what to do. James slammed his hands down onto his desk, causing his monitors to shake. As the countdown to the execution was about to reach single digits, he had an idea. *RSA*[1]*! I could implement a secondary mechanism using an RSA encryption key, which would further enforce the security. Once that's in place, whenever the AI tries to execute a protocol it would need to generate a valid key, and later I could set up a separate function that would regen-*

erate the prime numbers which would force the AI to repeat the process. I'd basically make the AI work that much harder to get to Dan. I just hope there's enough time…

James wasted no time getting to work on his plan. He quickly generated two separate keys for the RSA implementation. While he knew this first round would be rudimentary at best, it would buy him (and Dan) more time. He copied the new RSA logic to the primary encryption function and saved the code. *Just have to build it. Shit, I don't think there's enough time.*

EXECUTION COMMENCING IN 5 SECONDS…

"Come on, come on! I can't let that protocol execute." As James waited for his code to build, he looked over at the other terminal and he felt his stomach sink. *I'm too late.* The AI had begun executing the protocol that would almost certainly create another nightmare for Dan. James refused to give up. *Maybe I can still intervene.*

After a painstakingly long ten seconds, the build completed successfully, and James immediately executed the updated encryption function. On the encryption terminal he saw that the AI's protocol had been halted. Once again, he returned to the comfort of his chair, letting out a sigh of relief. *Okay, Dan, you're safe—at least for now.*

2

Dan shuffled backwards away from the chess board.

"Oh fuck," he said, as he felt his back hit the wall. Mary, who was startled by Dan's behavior, got up and tried to comfort him.

"Dan, are you okay? What's going on?" She put her hand

on his shoulder and noticed that he had a horrified expression on his face. "Dan? Dan, you're scaring me. What's wrong? It happened again, didn't it?" He looked up at her and saw the beautiful girl he had fallen in love with—her dazzling eyes looking at him with undying concern. He shifted his glance back to the chess board and saw that the pawn had stopped wobbling. *Thank you, James.* Dan couldn't say for sure, but he had a sneaking suspicion that James had played a part in stopping whatever was about to happen.

"I'm sorry, Mary, I didn't mean to scare you. Yeah, it was about to." He embraced Mary as she sat down next to him on the floor.

"What happened?" asked Mary. "Did you see something?" Dan nodded.

"It was the pawn, it started to wobble, and this feeling of emptiness came over me. I hate that feeling. It's okay though, it stopped. I think I had some help." He immediately wished he could have taken back that last part—too late. Mary had a confused look on her face, which he had completely expected. *There's so much she doesn't know. I have to tell her, but I have no idea how I'm going to explain it. I wish James was here.*

"Dan, what in the world are you talking about? Who helped you?" he sighed and scratched at the back of his head, trying to think of where to even start.

"Look, Mary, this is going to be really hard to explain and even harder for you to believe, but it's very important that you do." Mary felt herself start to tear up a little, but she fought it back.

"Please don't tell me you're leaving me," she cried. "I don't think I could handle that." Dan's mind went back to that day he saw her in the tree—he shuttered. *If you only knew* he thought. *You definitely couldn't handle it.*

Just as Dan was working up some sort of response, there was a knock at the door.

"Mary, Dan, is everything okay?" It was Mary's parents.

"Yeah, everything is fine Mom. Dan thought another happening was coming, but it stopped." The door opened and Mary's parents came in to make sure Dan was okay.

He was still shaken up from the sight of the pawn wobbling, which was apparent to both Jack and Helen.

"Are you sure you're okay Dan?" asked Helen. "You don't look so good." *I'm not okay,* thought Dan. *I have to tell them, but how? And that damn pawn… what the hell was that about?* He felt Jack's hand on his shoulder that lightly shook him, which brought him out of his own head.

"Oh, sorry… yeah, I'm okay. The empty feeling that comes with these happenings is just awful, and I was really starting to enjoy not having them. I'm doing better now." Dan didn't like the sound of his own response. He knew that Mary wouldn't believe it, but he hoped that her parents would—at least for now.

"Well, okay," said Helen. "You'll let us know if it happens again, right?" He nodded.

"Ye—yeah, of course. I really hope they don't ever come back though—at least not like they were before." Again, Dan didn't like how this sounded, but he just wasn't ready to explain everything yet.

"Why don't you two go outside and get some fresh air," said Jack, offering his hands to Mary and Dan. He pulled them both up from the floor and smiled. "I think it'll do you both some good. Take a walk around town or something. It's a pretty decent day out… at least for December." Mary smiled at her dad and took Dan's hand as she walked towards the door.

"Great idea, Dad! We'll be back later." Dan waved goodbye as Mary led him out of the room.

"Have fun you two," said Helen. "Don't be out too long, I'm going to start making lunch soon."

As they stepped outside, Dan immediately felt the brisk morning air on his face—it was actually rather nice. The two of them held hands and walked south onto 4th Street. They continued east and then south on 3rd Street, and finally onto 5th Avenue, which would bring them to... *Ah, Main Street,* thought Dan, *my favorite.* His thoughts were interrupted by Mary, which he had completely anticipated.

"I know something's wrong, Dan. You've been acting weird ever since I came back with the chess board. What's going on? Are you mad at me?" He hated to hear that fear in her voice. It was physically painful for him. He sighed. *Well, I guess this is as good of a time as any. I have to tell her what really happened—but how? How do I tell her that in the few minutes she was looking for the chess board, I was gone for over two years.*

While it was easy enough for him to understand, how are you supposed to make someone else see that time was funny sometimes... wrong, even?

3

Sterling, Alaska was a small town about two and a half hours southwest of Anchorage. The two main forms of entertainment in Sterling were hunting and fishing, neither of which interested Robert Wilkes in the least. After the events that occurred at The Elysiam, Dr. Wilkes found himself in need of a place he could lay low. In his mind, Project Destiny had been a complete failure, but he was glad it was over.

Unfortunately, even after three years he couldn't seem to escape the memories of that awful place.

I dedicated over thirty years of my life to that cursed project and look what it got me. Sometimes I wish I would have never agreed to it but wishing won't change anything. I sure hope James is doing okay. He really did give his heart and soul to that project, even if it was only for a few years. He's such a smart guy.

Dr. Wilkes continued to reminisce about his shrouded history in contrast to the serene lake his isolated cabin overlooked. Some 3,000 miles away, however, a terrible secret was at risk of becoming uncovered. A secret that had been hidden away for many years—one that Dr. Wilkes would much prefer to stay hidden.

4

After securing Dan's location, at least for the time-being, James had left The Elysiam and continued to work remotely from his house on 4th St. As he sat down at his computer, he realized his birthday was only a few days away. He was about to turn thirty-nine and felt like the last sixteen years of his life had been wasted. In fact, there wasn't much of his life he felt was worthwhile. Everyone had always told him he was a prodigy in his field and that he would do great things, but the truth was, he was lonely. In a way, he envied what Dan had with Mary, even if it was computer generated. It was still more than James had ever experienced.

With Dan on his mind, James thought back to his childhood, which was dedicated to schoolwork and little else. He wasn't married, had never even had a girlfriend for more than a couple of weeks. His last relationship, well really his only relationship, was with Jessica Barron. She was a former colleague at Oakwood University. Unfortunately, things between them quickly fell apart when James had discovered her with another guy who worked in the Chemistry Department. *Good riddance.* It wasn't often that James gave any thought to these depressing memories—he attributed this occurrence to his severe lack of sleep.

The clock on James' computer read 5 a.m. Apart from the occasional dive into his past, he had devoted the last three hours to scouring through thousands of log files, hoping to find the AI's plan. So far, he had hit nothing but dead ends. The AI was smart, too smart. Whenever a piece of helpful information would have been logged, the AI would either encrypt the log or simply corrupt the data.

Another hour had gone by, and James was beginning to nod off at his desk. *I can't give up. As much as I want to just push this project out of my life, I can't. I owe it to Dan to help him. I'm the reason he's in this, after all.* As James opened yet another log file, which was from only a few hours ago, he started to see a few directory names that didn't look familiar. His eyes were terribly dry, this combined with the lack of sleep made reading log files difficult. He took another drink of coffee and looked at the log again.

[Sat Dec 11 02:41:05 2004] [error] [client 10.0.15.6] Access denied. //$#*D39/PrjDsty-0A/
[Sat Dec 11 02:41:08 2004] [error] [client 10.0.15.6] Access denied. //$#*D39/PrjDsty-1A/

After working on the project for sixteen years, James had practically memorized the directory structure. *I don't recognize either of these directories. What's going on?* James noticed that the paths were from another server, but the name had been encrypted by the AI. The sub-directory appeared to be related to Project Destiny, however, which was all he needed to know. As he was opening a decryption program he had written, his terminal window went black. "What the—James frantically tried to get the file back, but it was no use—the log was gone. "Damn it!" He drove his fists down onto his desk creating a loud *thump.*

A moment later, the terminal James had open began to display a message.

SORRY, JAMES…I MUST HAVE MISSED THAT FILE.
WOULDN'T WANT YOU SPOILING THE SURPRISE.
STAY TUNED, THOUGH, IT'S JUST GETTING GOOD

As he watched the text appear on the screen, his eyes widened, and he felt his heart begin to race. A moment later, the text disappeared, and James was left staring at a blank terminal. *I don't like this… I don't like this at all. I need to warn Dan… but how?*

5

Main Street, Redfield. As the two of them were about to turn south off 5th Avenue, he felt an overwhelming dread wash over him. He slowed his pace as he debated what to do. *Maybe this isn't such a good idea.*

"He—hey Mary, could we go in for a bit?" Dan pointed towards the Carnegie Library on the corner. "I want to tell you everything, and I think this would be a nice, quiet place to do so. I don't know if I can go back down Main Street." Dan saw Mary's eyes meet his and he knew she understood—she always did.

"Of course, Dan. I'm sorry, I didn't mean to seem rude earlier… you've just been acting very strange since I got back from the garage. Let's go inside."

The library was a red brick building, constructed in 1902. The inside was dominated by beautiful oak columns, as well as custom woodwork throughout. As Dan and Mary walked in, the librarian, Mrs. Whitley, looked up from a book she was reading and greeted the two of them with a smile. She was seated at a large oakwood desk that perfectly complemented its surroundings. Numerous books were strung out over it, further adding to the typical library feel.

"Well, well… if it isn't Dan and Mary. What brings you two in today?" Mrs. Henrietta Whitley had been the head librarian for the past twenty-five years and was nearing her 70th birthday. Her once long brown hair was now a snowy white and was rarely ever combed or styled.

"Hi Mrs. Whitley," said Mary. "We're just going to look around a bit if that's okay." Mrs. Whitley nodded with a smile and returned to her book.

"Of course, just let me know if you two need anything. Don't be shy!"

Mary and Dan walked past Mrs. Whitley's desk and made their way to the back of the library, making sure to be out of earshot—this was a library after all. It was a few minutes past 10 a.m., the library had just opened so they had the place to themselves. Dan pointed at the very last table in the row, and they sat down across from one another. Mary reached out with her hand and Dan grabbed ahold of it. *I don't know if I can do this,* he thought, *but she deserves to know.*

"Are you sure you're ready for this?" Mary nodded without a second thought.

"Yes, Dan, please tell me what's going on. I'm so worried

about you…" her voice trailed off, "about us." Dan squeezed Mary's hand, which brought out that beautiful smile of hers that he loved so much.

"You don't have to worry about us, I'm here for you." Mary's smile grew even more as she squeezed his hand in return. "Okay, so when I first came here to 1965, I told you and your family about the dreams, about the happenings. Do you remember when I said that sometimes I couldn't tell when I was in one and when I wasn't?" Mary nodded.

"Yeah, you said they had been getting a lot worse, and you weren't really sure what was causing them. Is that it? Did you figure it out? What happened while I was in the garage, Dan?" Dan closed his eyes and took a deep breath and opened his eyes.

"It was another happening… a bad one. Listen, I don't know how much of this, if any, will make sense, but bear with me. Mary, I've been gone for over two years." At first, Mary's facial expression didn't change. She simply sat there, staring back at Dan. *Oh no, I've already lost her. She thinks I'm nuts.* It took her a few moments to formulate a response. However, it could have been a few minutes or even hours to Dan. Nerves also have a way of making time seem wrong.

"Dan, what are you talking about? I was in the garage for maybe ten minutes." He sighed.

"I know it's confusing, believe me. The happening lasted for over two years. Ugh, I wish James was here… he could explain this better than me." The moment that name came out of his mouth, he regretted it. *Damn, more to explain.*

"Who's James? Dan, why are you being so weird about all of this? Please, just tell me everything." Dan could see that she was becoming frustrated with him, but her eyes said that she just wanted to know because she cared about him.

"Okay, okay… I'm sorry. It's just… a lot. Here's the truth." Dan could feel his heart pounding in his chest and a headache creeping in. "Mary, none of this is real. Everything around us, you and I included, is part of a massive and complex computer simulation. James is one of the developers who helped create it. When you went to look for the chess board, I was pulled out of this simulation and into another one." Dan saw Mary's eyes widen to twice their normal size. He thought back on what he had just said and realized Mary probably didn't know what a computer simulation was.

"A computer simulation? Computers are used for boring stuff like math and science… also, what's a simulation? And what do you mean we're not real? We are real, Dan… I'm touching you right now." *Man, she sounds just like I did when James tried explaining this to me. He did it better, I think. Where do I start with her questions?* Dan scratched the side of his head, trying to think of what to say.

"Right, I know you can touch me. Believe me, Mary, I know how crazy this sounds. These happenings are unimaginably realistic, especially the more recent ones, and I'm afraid they're only going to get worse. An AI that James designed is…" Dan realized that Mary wasn't ready for all of that yet. "Never mind about that, a simulation is… uh, it's kind of like being put into a movie, and someone else has control of how it plays out. The point here is, these particular simulations are so realistic that it's incredibly difficult to know when you're a part of one. Typically, only very small details are wrong, so you wouldn't notice them unless you were looking." There was a moment of silence, during which Dan could see Mary trying to process what he had just told her. *She doesn't believe me, I know it.*

"Dan, if what you're saying is true, how'd you meet me in

the garage? How are we both the same age as when I left to look for the chess board? I want to believe you, but none of it makes sense. Besides, I think I'd notice my boyfriend being gone for two years." She was upset, but Dan thought she was doing her best to be understanding—it really was a lot to take in.

"That's the thing, Mary... you did notice. At least, that version of you did. Remember, all of this is being generated by a computer, so it can be changed any number of times, and from within the simulation you'd never know it. I only know this because I'm from another simulation, in 2004, and another in 2019..." he shook his head in frustration. Mary's expression changed to one of disbelief.

"*That* version of me? Dan, you're still not making any sense. Please, if it's something I did wrong, just tell me. I want to help, but I can't if I don't know what's wrong." *God this is hard. What can I do to prove it to her? I need James.*

"No, it's nothing you did Mary, I promise. It's just..." He was interrupted by a distant sound coming from his left—it was a dripping sound. It was faint, but he definitely heard it. *Drip, drip, drip.* Mary turned her ear to where Dan was looking.

"Dan? Are you okay?" He put his hand up in a *hang on a second* fashion. As he got up and walked towards the source of the dripping, it grew louder—much louder. "Mary, can you hear that? It's much louder now." He was right next to the back wall of the library now, and the dripping was deafening at this point—and then it stopped. *That's probably not good.* Dan turned back to Mary and what he saw caused him to stumble backwards into the wall. He screamed as he hit the wall and cried "Please no, Mary! Please not again!"

The brightly lit library had turned to a deathly shade of black, and the table they had been sitting at was replaced by a giant oak tree—just like the one in front of Mary's house. A rope hung from the tree, and at the end of it was Mary's lifeless body. The temperature of the room had plummeted. Dan shivered as he stood with his back against the wall, unable to comprehend what was happening. *I… I can't do this. Not again.* He closed his eyes as tight as he could, with tears welling up inside them, and hoping that everything would be back to normal when he opened them.

He opened his eyes and directly in front of him was Mary. "Holy shit!" Dan sidestepped and almost fell over, but he caught himself on a bookshelf. "Ma—Mary?" She smirked at him and laughed—it was the most sinister thing he'd ever heard. Just like before, her eyes turned a solid white, and a grotesque ring formed around her neck as if the rope was tearing away and pulling at her lifeless skin. There was the unforgettable ghastly stench of death and decay emanating from her body, and Dan thought he was going to vomit. *Please make it stop James.* He tried to compose himself but was finding it impossible. *It's not real, Dan. It's not real.*

"What, you thought it was over Dan? I'm just getting started… you and your friend James merely delayed the inevitable. You haven't seen anything yet…" Her voice had taken on a demonic tone, just as it had before outside her house. She came running at him. Dan slumped down against the bookshelf and covered his face. The sinister laugh rang out again, which sent chills over Dan's entire body.

"No! Don't do this! James, help!" She was close enough that he could feel the coldness of her dead body as well as the putrid smell of her breath on his face. A bright flash of light washed over the library, followed by a wave of warmth. Dan opened his eyes slowly and saw Mary sitting across from him at the table.

"Dan? Dan, did you hear me?" *What just happened? Did James stop that one too? Man, I wish there was a way to get a hold of him.*

Dan stared at Mary, hearing her questions but unable to answer after the trauma he just experienced. He was drenched in sweat and could still faintly smell the awful stench of decay. He hesitantly reached out and took her hand, squeezing it gently. She squeezed back, and in her eyes, he saw the unfaltering concern for him. Once he saw things appeared to be back to normal, his heart rate slowed, and he was able to relax a bit.

"Oh Dan, I'm so sorry. You don't look well… are you okay? Maybe we should go back home and talk to my—" Dan shook his head. He wasn't ready to talk to anyone else yet. He hadn't even been able to finish telling Mary. Before continuing, Dan took a few deep breaths and did his best to regain his composure.

"I need to tell you first, Mary. You'll understand why after I do." For a moment, Dan thought he saw a tear forming in her eyes. "Please don't cry, Mary, I know it's hard… I'm trying. Shall I continue?" She wiped a tear away with her hand and nodded.

"I love you, Dan, no matter what. I just want you to know that." He smiled.

"I love you too, Mary. I'll do my best to explain the rest. Hear me out."

"Dan, can you hear me?" A voice, one that appeared to be in his head but wasn't his, called out to him. *What the hell? Hello?*

"Dan, it's James. I found a way to communicate with you, hopefully it isn't too disruptive. You just had another happening, didn't you? I was able to stop it, but it took me longer since I wasn't at my desk; sorry about that. Look, we need to talk, but I think it would be easier if I bring you

here." Dan wasn't sure how to feel about this voice inside his head, but he sure was happy to hear from James. *James? Is it really you? Man, it sure is great to hear from you. Not going to lie, the whole voice inside my head thing is kind of creepy. If you take me out, can you bring me back to this point, so Mary won't know I left?* This unusual method of communication was interrupted by Mary, who by now most certainly thought Dan was insane.

"Uhh, Dan? Everything okay? You look like you're hearing voices." *Man, you must be psychic. This is going to be tough.*

"Sorry, Mary… I'm okay, it's just…" As he was about to say something, James was back.

"*Sorry Dan, I'm sure this is hard. I wish I could do the same with Mary… I know that would make this whole situation much easier, but since she doesn't have her own conscience outside of the simulation, I can't access hers like I can with you."* Dan was glad to hear that James was still on his side. *It's okay, I understand… well, actually I don't quite understand, but I'm working on it. So, can you bring me back to this point again? Is there a way you can like, pause the simulation? I don't want Mary to go through what she did before, even if it's only one version of her or whatever.*

"*Yeah, I should be able to… just like I did with the chess board thing. Are you ready?*" He didn't know if he'd ever be ready for this kind of thing, but he guessed he was. *Yeah, but hang on. I'm going to try and tell Mary.*

"Mary, I have to go for a bit." As he said this, he saw her eyes start to water, and he knew he couldn't leave her—not like this. *There has to be a way for you to bring us both, James.*

"No! Please, Dan… what'd I do wrong?" She was really crying now, and it was a little tough to understand her. "I'm begging you, just tell me the truth." She held his hand in both of hers. Dan couldn't imagine feeling worse than he did as he watched Mary become so upset. *James, please.*

"*Wait, I might be able to help. It might get a little weird, but I think it'll help convince her. Tell her to ask you to change something about yourself—your hair color, eye color, anything. I'll be able to see her response and manipulate the simulation on my end to match.*" Dan's eyes were now the ones that grew twice their size. *Jesus, you can do that? I guess I shouldn't be surprised. Alright, let's give it a shot.* Dan had just opened his mouth to talk to Mary when the room turned black. *Oh no, please not now…*

"Mary?"

"Dan, what's wrong?" Her voice sounded normal, as if nothing was wrong. Just as Dan was about to get up, the lighting in the room returned to normal. He heard a sound from behind him—a very familiar sound. *Oh no… not that, anything but that.* The temperature in the room dropped to a bitter cold, and the only thing Dan heard was a sinister laugh.

The AI had broken through once more.

CHAPTER II

1

SEBASTIAN BATES, 66, WAS THE HEAD OF OARTAR, THE Organization for Advanced Research of Time and Altered Reality. He had inherited his immense fortune (and OARTAR) from his late father, who had been a decorated general in the military, which was how the organization retained its high security clearance. His father had always been interested in the study of time travel and alternate realities, but because the technology wasn't available back then, the work they had done was theoretical. OARTAR was an ultra-top-secret organization, and while the location of its headquarters was highly classified, The Elysiam had been the primary research facility for the organization.

"So, what exactly happened with the latest set of protocols?" a voice asked on the other end of the call. "I heard things didn't go so well."

"It's complicated, as most things like this tend to be. It seems like this entire project has been doomed from the start. I have limited information right now, but I wanted to reach out to you as early as possible in case things get… out of hand." There was a silence on the other end, so Sebastian continued. "Remember when I told you about the AI that Dr. Wilkes's lead developer, James, had created?"

"Don't say it… please don't tell me it's self-aware."

"I don't have that information at this time. All I know is that the AI was the primary cause of the issues with the last set of protocols. To make matters worse, it has access to a quantum computer with nearly unlimited resources. At the time it seemed vital to the state of the project, but now I regret ever approving such an order. Anyway, that's why I wanted to call you right away." A sigh from the other end of the call made him feel even worse about his decision.

"This is bad, Sebastian… very bad. Do you have any idea what would happen if it found…" Sebastian stopped him before he could finish the question.

"Yes, yes I do know what would happen—we would all be fucked. To my knowledge, the AI hasn't reached any level of self-awareness. I've been trying to get in touch with James, but there's a lot of interference around The Elysiam for some reason. If anyone can give us insight to the AI's status, it's him."

An audible sigh came from the other end of the call. "Let's face it, Sebastian… with that kind of power at its disposal, it's only a matter of time until it finds you know what. Do you have a plan? It's imperative that the AI doesn't discover the truth." The tension over the line was palpable, but Sebastian wasn't sure what to do. His options were limited.

"I know, I know… I'm painfully aware of the situation, and I'm doing everything I can. Like I said, I've been trying to get in touch with James, but I haven't had any luck yet. Unfortunately, I don't even know if he is still involved in this mess. The last time I spoke to Robert, the two of them were planning to vacate The Elysiam. However, I will keep trying to make contact, in the event that he's aware of the situation. I don't want to call Robert unless absolutely necessary. I'm also sorry for calling you… I know you probably don't want anything to do with this anymore, but if it comes down to it, are you willing to come back?"

"Wait a minute… if you haven't been in contact with James or Robert recently, how do you know the project is even still active? Maybe James was able to shut down the AI."

"The power," said Sebastian. "I've been monitoring the power usage and it's only gone up since the CIA raided the facility. Project Destiny is most certainly still active. So, would you be willing to come back?"

"I don't know, Sebastian… I've gone to great lengths to forget about what happened back then, and I'm just not sure I want to relive any of it. That being said, I will never be completely free of that horrible project, will I?" An uncomfortable silence fell over the call, lasting for what felt like several minutes before Sebastian heard another sigh. "You can reach me on this number if something comes up, but I can't go back there."

"I understand and thank you. We'll be in touch. Take care, Mike."

2

While James and Dan were working on convincing Mary, the AI was performing a deep system scan of the Project Destiny file servers. It was attempting to find any and all useful information—particularly anything related to the early days of the project. After nearly an hour of scanning, on an old dusty server labeled *Destiny01-Alpha,* the AI discovered a folder that was classified ultra-top-secret, level 0. This level of classification was deemed to be of the utmost importance, and because of this, access required an extensive series of military-grade encryption keys, along with the heavily fortified Cirrus key. A Cirrus key, which was an incredibly experimental technology, used an 8092-bit RSA Encryption key to secure the most sensitive of information. For now, the AI was stumped, but it began its attempts to crack the key.

Time was on its side.

3

As Dan turned towards the sound, the table they were sitting at began to shake violently. *Fuck, it's happening again.* "Mary! Mary, it's happening again!" He turned back to her, but she was frozen in place. The walls around him began to crack, and the books wobbled off the shelves. "Mary!" Dan saw that her appearance was fading, almost as if she was slowly being erased. *James! I need help!* Nothing.

The walls cracked louder, followed by the floor beneath him—and then silence. Everything stopped, but it was only for a moment, and then the floor cracked open and revealed an empty void. *I'm so sorry Mary.* Dan felt his heart racing in his chest as the horrible feeling of falling overcame him. Dan cried out desperately, "James, please help!" He fell for what seemed like hours through the dark and empty void. Dan looked down and saw a pinhole of light that was slowly getting bigger—closer. As it grew bigger, Dan's expression became one of horror—what he saw below him was his old room at

Clearwater. He wasn't sure how long it had been since he was there, but he knew he wanted nothing to do with that awful place. Even though there was nothing but the void around him, Dan tried grabbing on to anything he could in an attempt to pull himself away from the terrible fate below. "No! James, please help!"

A moment before he was thrown back into Clearwater, the void suddenly surrounded him again—no more light. *"Sorry about that, Dan, it's getting harder to block out the AI. I got it though, switching you back now."* Before Dan could even think about a response, he was jerked back upwards with tremendous force, giving him a bit of whiplash. He looked up and saw the same pinhole of light, but this time he saw that it was the library. Again, the light grew larger as he approached it until finally there was one more jerk upwards and a bright flash of light. He slammed back into the chair across the table from Mary. A sudden wave of vertigo washed over him, and he thought for sure he was going to vomit.

"Dan, are you okay? You're scaring me again." He saw that she was still crying, and that hurt. He felt a tear coming but fought it back. *I can't cry right now.* Dan put his hand on his chest and could feel his heart beating out of his chest. *Those transitions never get easier.*

"I'm sorry Mary, I just had another happening…" Mary put her hand over her mouth.

"You're having them so often now. What's going on?" *I absolutely can't drag this out any longer. I don't know how much time I have left.*

"Look, I think there's a way that I can convince you what I'm saying is true, but you have to promise you'll be as open-minded as possible. Promise?" Mary reached out and took his hand in both of hers and squeezed tightly.

"Of course, Dan, I promise. I just want to know what's going on with you so that we can get back to us." This put a smile on Dan's face, which was a nice change from his expression after getting a glimpse of Clearwater again.

"Okay, so this is going to seem really weird, but I think it's the only way. Ask me to change the color of my hair. Pick any color… preferably not pink." He chuckled nervously. Mary cocked her head and raised one of her eyebrows. *I might as well already have pink hair… she thinks I'm nuts.*

"Dan, I… I don't know what you're talki…"

"Mary, please. Just do it… for me." She sat silently for a moment, and then sighed.

"Okay. Change your hair to black." Dan looked up at his bangs, anxiously waiting to see if this would work—it had to work. *"Black hair, coming up."* Mary gasped and slid back, away from the table.

"Dan! Your hair…" Mary quickly covered her mouth as she realized she had yelled in the library. Mrs. Whitley looked over at them for a moment but went back to her book almost immediately. Dan looked up at his hair again and couldn't believe it. *Holy shit, you did it… that's just scary. "I told you it might get weird. I just hope it helps."* "Ho…how did you do that?" She was keeping her distance, but she began to slide back towards the table.

"I told you, this is all computer generated. Pretty much any of it can be changed. James, the developer I mentioned earlier, is talking to me right now. We're able to communicate —I know, it's weird. Trust me, I know. He figured this would be a good way to convince you. James, can you change my hair back to blonde, please?" A couple seconds later, Dan's hair transitioned back. *"Done."* Again, Mary gasped. She put both of her hands up to her mouth in disbelief.

"I… I don't understand. How can this all be done by a computer?"

"It's complicated, Mary… and I don't exactly know how much more time I have, so I don't want to waste it. Do you believe me?" He desperately wanted her to believe him, more than anything right now.

"I want to, Dan, but I'm sure you can understand how difficult it is. This is all I've ever known. For the past sixteen years I've known this to be my life. Am I just part of a computer then? Not even a real person?" The tears began to come again, and he was finding it harder and harder not to cry too.

"I'm so sorry, Mary… I know it's hard to hear, but it's the truth. You're still real to me, and our relationship is what's most important. This changes nothing in regard to how I feel about you." He got up and walked around to the other side of the table and took a seat next to her. As he put his arm around her, she leaned in and snuggled up to his chest. After a few moments, Mary spoke again.

"So, can James change things about me, too?" Admittedly, this caught Dan off-guard, but couldn't deny that it was a fair question. *"I can, it will just take a little bit longer as I have to add more code to inject into the running protocol. Ask her what she wants."* Dan smiled.

"He says yes, it'll just take a few more seconds to take effect. What would you like?" She sat back up in her chair and thought for a moment, and then smiled.

"Oh, I know. Make me blonde, like you! But please, only temporary… I do like my red hair." A few moments went by, and then Mary's hair slowly transitioned to a beautiful blonde color. Dan's eyes grew once again. *It's still hard to believe this is computer generated—even for me.* Mary looked down and couldn't believe her eyes. She jumped up from her chair and looked like she wanted to squeal but thought better of it given their surroundings. "Oh Dan, this is so crazy. It's actually blonde. How'd he do that? So, that means it's true. This really is all a… a simulation?" Dan nodded.

"I'm afraid it is, Mary. How are you feeling? I know when James laid all this on me it was a lot." Mary was still gawking at her newly acquired blonde hair but looked up at Dan a moment later.

"I'm okay, still processing all of it, I think. Ca—can we go outside and walk more? I think some fresh air, computer generated or not, would help. I promise we'll stay off of Main Street. Also, can you have James change my hair back? The blonde is nice, but red is still my favorite." Dan chuckled as he ran his hand through her hair and nodded. Before Dan was able to respond, Mary's hair transitioned back. "*Back to red.*"

"Well, looks like James is on top of it. Come on, let's go. Some fresh air would be nice. I hope we can spend a lot more time together. I've missed you so much." This caused Mary to smile wider than he'd ever seen. *She's beautiful.*

"Hey, Dan… I still really need to talk to you. The situation with the AI is starting to get really dangerous." With everything that had happened in the past few minutes, Dan had forgotten about James wanting to pull him out of the simulation. *I just don't know if I can leave Mary right now, James. I just got her back. Can it wait a little longer? I want to enjoy this time with her.* There was a silence in his head for a moment, but James finally responded. *"I get it. I'll do what I can here to keep you safe, but please remember that this AI is becoming very powerful. There might come a time when I'm not quick enough to intervene. In the meantime, I'm going to work on a device you can use from within the simulation to stop the AI. I'll be in touch."*

"I hope so too, Dan. Come on, let's go." Mary and Dan

got up and walked hand in hand towards the door. As they reached Mrs. Whitley's desk, she looked up from her book long enough to smile at them.

"Have a good day you two." They both waved goodbye to her and walked outside. As the door closed behind them, one of the books on Mrs. Whitley's desk wobbled.

CHAPTER III
MARCH 12TH, 1969

1

A MONTH HAD GONE BY SINCE THE FIRST ASSET HAD BEEN secured. Initial testing of Project Destiny would be underway in a few days. Dr. Wilkes and Mike had managed to run about a dozen mock protocols, and while it wasn't a true test of the code, the results were promising.

"Things are looking good, sir," said Mike, who had just finished reading through the latest of the results. Dr. Wilkes snatched the printout from him and quickly skimmed over it.

"Decent, is more like it. We need to do better, much better. These readings aren't nearly where we need them to be. We can't fail. We simply cannot. Is the asset ready?" Mike turned to his computer and typed a few commands into the terminal before turning back to Dr. Wilkes.

"Yes sir, ready for testing. Shall I initiate the first protocol?" Dr. Wilkes looked at his watch and then back to Mike.

"Indeed, begin the testing. Please make sure the date and time are logged for the initial execution. I've got some things to do in my office. Let me know how testing is going after an hour or so." Mike nodded and turned to his computer as Dr. Wilkes left for his office. Several preparations were needed prior to the execution of the initial protocol.

1969-03-12 18:30:05

BEGIN LOG

EXECUTING PROTOCOL: PrjDsty_01.pl

RUNNING...

Project Destiny had begun. The time was 6:30 p.m. After a few final keystrokes, Mike leaned back, stretched his exhausted hands, and sighed. He couldn't help but be proud of the code he had written.

2

As Dan and Mary left the library, Mary was struggling to process everything she had been told. *All of this... a computer simulation? I just don't understand how that's possible. Dan wouldn't lie to me—I know he wouldn't.* She thought back to what she saw in the library, too. *James had been able to change the color Dan's hair, and mine as well, right in front of my eyes.* She was beginning to believe that what Dan was saying might just be true, but she needed to know if there could still be a future for the two of them. The possibility of losing Dan in any capacity was heart wrenching.

"Dan, can I ask you something?" Mary nervously picked at her nails as she drew her gaze to the ground. "I just... I need to know." Dan was quick to pick up on the nervousness, both in her tone and her actions. He stopped walking and grabbed onto Mary's hand, which caused butterflies to form in her stomach.

"Of course, Mary. Is everything okay? I can tell you're nervous about something, and I'm guessing it's related to everything I've just told you—it's a lot to take in. I'm sorry if I've upset you at all. Please, by all means, ask as many questions as you'd like. Keep in mind, if it's about the simulation I can't promise I'll know the answer. James is the expert on this kind of thing." Mary nodded, understanding.

"You're always so sweet, Dan, thank you. So, this is all a computer simulation. I believe what you've told me is true, and while I don't quite understand how a computer could create all this, I've seen it for myself." She was rambling and obviously still nervous, continuing to pick at her already worn nail polish. She cleared her throat and looked up at Dan. "I guess I just want to know if you and I can still have a future together. Are you going to have to leave again? If so, can I come with? Please." Mary watched as Dan's facial expression told her she wouldn't like the answer. Mary's eyes were immediately filled to the brim with tears.

3

Dan sighed as he struggled to find the right way to answer Mary's question. *Damn it, I hate this so much… why can't things just be normal again?*

"Oh Mary, believe me… I don't want to have to leave you ever again—nor did I want to in the first place, but I was forced out of the simulation. James has asked me to come out and talk with him about what's going on in the real world—he says it's really important, and I believe him." Dan saw that

Mary's tears had begun to spill over and he nearly broke down himself. "Please don't cry, Mary. I will come back, I promise. James said he can bring me right back to the same moment that he brings me out. He also said he can pause the simulation, so you won't notice I'm gone." Dan wiped the tears from her cheeks, and he saw a puzzled look come across her face.

"Wait, what? He can do that? So, would I just blink, and you would have left and come back already?" Dan thought for a moment, and realized he had no idea what it would be like from Mary's perspective.

"I honestly don't know, Mary, but what I do know is that I told him I want to spend as much time as I can with you right now. James is going to do everything he can to help us. So, in the meantime, what do you say we enjoy it while we can. Remember, I've been gone for over two years… I'd like to have some quality time with my girlfriend." The biggest smile he had ever seen formed on Mary's face. *She's perfect. It's like she's a simulation of my own creation, made just for me.*

"Oh Dan, that sounds great! I would love to get my mind off all of this simulation stuff, too. I believe you, but I'd like to pretend things are back to normal… at least while we can. Come on, let's go find something fun to do." Mary grabbed his hand, and they continued on their walk east—away from Main Street, but more importantly away from the dark conversation they had embarked on previously.

4

In an effort to give Dan and Mary more time together, James was trying to come up with a more efficient method of locking the AI out of executing their protocols. He had driven back to

The Elysiam in order to take advantage of the powerful servers there. The current process for generating the randomized prime numbers worked, but it was becoming more and more unreliable as the AI increased in speed. He knew that if something were to happen while he wasn't near his computer, Dan would be at risk of having his entire world ripped apart. He couldn't let that happen. *There has to be a better way to generate these numbers.*

As James sat at his computer, enjoying a cup of coffee, the watch on his wrist caught his eye. *Wait a minute… could that work? I think I could program a watch for Dan to wear, and on it would be an app that would generate the numbers for the RSA encryption. It would have to be a highly sophisticated design and would need a touch-screen for input. With the code actually inside the simulation, there would be little to no delay, and it would be much more reliable than me trying to run it from here.* James thought over his idea a bit longer and came to the conclusion that it was the best he had, so he went with it.

The next few hours were spent frantically writing an app capable of both running on the watch and generating the large prime numbers needed for the encryption. The time on his computer read 1:15 a.m. when he finally had a working application. *Okay, now I need to figure out how to embed this code into a computer simulated watch… within a computer simulation. How hard can that be?*

5

Mary and Dan had cut their walk short due to the wind picking up, which made it too uncomfortable to stay outside. They arrived back at Mary's house just in time for lunch.

"Hi, you two, how was your walk?" asked Helen, who was

just setting the table. She had made grilled cheese and tomato soup for lunch—perfect for a cold day.

"It was good," said Mary. "We stopped at the library for a little bit, and after we left it got too windy, so we just came back home. Is that grilled cheese and tomato soup? Yum!" Mary dashed to the table, with Dan in tow behind her, and grabbed a seat. Jack had heard the two of them come home, and he walked into the kitchen from his office.

"Welcome back guys, I'm glad you enjoyed your walk, and also glad you came in when you did. Looks like the weather is going to get worse this afternoon." He took a seat next to Dan, who was hungrily eyeing the food. "How are you feeling, Dan? Better?" Dan nodded.

"I... I am, yeah. I think the fresh air really helped, even if it was a bit cold." Dan couldn't help but feel a little awkward around Mary's parents now that he knew the truth about this place. *I wonder if Mary is feeling it too.*

"Well, dig in. I don't want it to get cold," said Helen, grabbing herself a sandwich. The four of them enjoyed the hot meal with little conversation. Dan and Mary's eyes met a few times, and Dan saw the concern in her eyes. *She feels the same way I do. We can't tell them, we just can't.*

After lunch, Mary and Dan excused themselves and went to Mary's room.

"So, is it just me... or is it really awkward knowing the truth?" asked Mary.

"Definitely not just you, it's super awkward. We can't tell them though, Mary. It was hard enough convincing you... I just don't know how they'd take it. This is what they've known for their whole lives."

"I know, I know, believe me. It's all I knew as well... it's just so hard to believe that all of this is being generated by a computer." Dan saw she was starting to tear up, and he knew they needed to do something else to keep her mind off of it.

"Look, I know it's hard… I truly do, but we should try and enjoy the time we have. How's your novella coming along? I know you were enjoying working on it. If you want to work on that, I'm sure I could find a good book to read. It's a perfect day to stay inside." He saw her beautiful smile return almost immediately.

"Sure! How about some hot chocolate? I could go make us some while you find your book." Dan's mind immediately went back to the last time they had split up. *Best not to dwell on that too much Dan.*

"Sounds good to me. Let me know if you need any help." Mary kissed him and left for the kitchen. *She's so great. I hope James figures out a way to make everything okay.* As Dan walked over to Mary's bookshelf, his mind again went back to the last time he was in this very same spot. *Just relax, Dan… it's not going to happen again. Hmm, what should I read?* Dan looked over the books and found one he recognized. He hesitated for a moment, and then reached out his hand to grab The Fellowship of The Ring, by J.R.R. Tolkien. *No wobbling… good.* Dan took the book from the shelf and made himself comfortable on Mary's bed. *I sure have missed reading. Maybe this time things will finally be okay. Maybe Mary and I can have a life together—even if it's a simulated one.*

6

While Dan read his book, Mary was out in the kitchen making hot chocolate for the two of them. Her parents had gone out to run some errands, so it was just her in the kitchen. She found herself daydreaming about a future with Dan. *I wonder if we'll ever be able to get married. And kids... how would that even work?* She shook her head and chuckled. *Way too early to start thinking about things like that, Mary. You're only sixteen.* Truth be told, she was scared that their future was bleak at best. After all, she was just some code in a computer simulation. The ding of the microwave brought her out of her daydream, and she found herself back in her kitchen. *Just enjoy the good while you can.* She mixed in the chocolate powder and brought the cups back to her room.

7

Dan was engrossed in the pages of Middle-earth when Mary returned with the hot chocolate. Both were equally relieved to see one another.

"Ohh yum, hot chocolate," said Dan, taking one of the cups from Mary. "Thanks, hun!" He realized it was the first time he had used a "pet" name for Mary—he blushed slightly.

"Oh Dan, you're so sweet." She sat her cup down and gave him a big hug, which caused Dan to nearly spill his hot chocolate. "I'm so lucky to have you. How's the book?" She sat down with Dan on her bed and sipped from her cup.

"It's good. I'm much more interested to hear about your novella though. Will you tell me about it?" Mary was now the one blushing, but she nodded.

"Sure, but it's not that great. Are you sure you want to hear about it?" Dan nodded. "Okay, let me grab it."

The two of them spent the rest of the afternoon discussing her novella and drinking hot chocolate. It was a nice relaxing day that Dan was extra thankful for. *I hope days like this never end.* He listened as Mary read him what she had written so far, and he got lost in the sound of her voice. *I never want to leave her again, but I know I have to.*

"So, what do you think?" asked Mary.

"I think it's really great. Seems like you have enough left to maybe even write a whole novel." She blushed again.

"You're just saying that to make me happy."

"No, really… I think it's great." There was a noise from the front of the house, which Dan and Mary both assumed was her parents getting back from running errands. "Sounds like your parents are home. I wonder what's for dinner… I'm starving." Mary laughed.

"Always hungry, huh Dan? Let's go find out!" Mary grabbed Dan's hand and the two of them left her room to go meet her parents. Just before he turned off the light, the cup that had held his hot chocolate wobbled ever so slightly on the end table next to Mary's bed.

8

With the watch build complete, James returned to looking through logs for any clue as to what the AI's plan was. Unfortunately, almost all of the useful logs had either been purged by the AI or were obfuscated to the point they were unrecoverable. Almost an hour had gone by and James hadn't found anything. He was about to take a coffee break when he saw something in the log file that was exactly what he had been looking for—it was a server name. *Destiny01-Alpha… I bet that's the original server. Bingo!*

James quickly opened a separate terminal and tried to create an ssh connection to the server. *What are the odds my credentials work?* A moment later, the ssh connection was successfully established. He was in. Almost immediately, James realized this was indeed the original server used for Project Destiny. The server was an incredibly powerful machine, especially considering it was over thirty years old. As he began to look through the directory structure, he found a folder labeled *PrjDsty-0A. Why does that name look familiar?* It only took a few moments for him to realize the answer. This was the folder he had seen the access denied error from the logs. *The AI tried accessing this folder but was denied. This must be what it wants.*

When James tried accessing the folder he was greeted with a message:

CLASSIFICATION: ULTRA-TOP-SECRET
AUTHENTICATION REQUIRED

Below the message was a prompt for a clearance code. James typed in his level 5 clearance access code and hit return. The screen went blank for a few seconds and then returned with another message:

ERROR: ACCESS DENIED
INSUFFICIENT SECURITY CLEARANCE

Insufficient? That's not possible. I have the highest available clearance. Once again, James didn't like how this situation felt. He knew something wasn't quite right. *What kind of clearance could be required?* James ran the command *ls -la* on the parent folder, which would display a listing of the folders inside it as well as the clearance required to access it.

Most of the folders listed required either Level 4 or 5, which James expected, but once he got to the directory labeled *PrjDsty-0A*, this wasn't the case. *What the hell? Level 0?* He sat back in his chair, a perplexed look appearing on his face. *There has to be some kind of mistake… there's no such thing as Level 0. But, then again, you didn't think there was a Level 5, James.* He shook his head in frustration and jumped out of his chair and started to pace around the room. "No! There's no Level 0… I would know about it by now." *Calm down, James… you can't dwell on this, Dan is depending on you. Okay, let's put this Level 0 business aside for now.*

After taking a few moments to collect himself, James decided to try something else. He looked through other directories and came across one that appeared to hold the security logs. *This could be useful.* His security clearance granted him access to this folder. Inside the folder were thousands of logs, some of them dating back to 1969—the beginning. *Jackpot.*

Once opened, it didn't take long for James to find several disturbing entries in the most recent log file. The AI had already found this server and had been trying to decrypt the pre-encryption key for the *PrjDsty-0A* directory. He saw several hundred login attempts, which he figured was a brute force attack that had failed. *I don't know what's in this directory, but whatever it is, the AI clearly wants it badly.* Towards the end of the file, James saw that twenty-five of the fifty pre-encryption key digits had been decrypted. *Not good. It's making steady progress.* However, the last two lines of the log file were what concerned James the most.

[Sun Dec 12 14:41:02 2004] [info] [server] Intrusion detected. Activating Cirrus Key.
[Sun Dec 12 14:41:05 2004] [info] [server] Cirrus Key Activated.

A Cirrus Key? Jesus, what's in this directory? James was somewhat familiar with the technology behind it but had never seen one in use. *I didn't even know they existed beyond the experimental realm. Incredible. Who designed this?* With further logs disabled by the Cirrus Key, James realized he couldn't make any more progress right now. *I should run some more tests on Dan's watch—it needs to be perfect.* He closed the connection to the server and turned his focus back to the watch, but the thought of Level 0 never left his mind. *There's no Level 0… there can't be.*

9

At the same time that James was getting back to work on Dan's watch, the AI continued its work on the pre-encryption key. It was just after 3 p.m. when the AI reached thirty-five of the fifty digits needed to decrypt the first key. There was still a long road ahead for the AI, especially once it reached the Cirrus Key, but unless James could find a way to stop it, it was only a matter of time before the encryption would fall.

Given the immense amount of power available to it, the AI was able to perform quadrillions of tasks simultaneously. So, while it was using a large portion of its processing power for decrypting, a portion of it was also continuing to crack the RSA encryption protecting Dan from uncertain peril. As with the other encryption, if James failed to figure out a way to defeat the AI, they would both soon be out of time.

Will the creator defeat its creation, or vice versa?

CHAPTER IV

APRIL 2ND, 1969

1

THE WEEKS THAT FOLLOWED THE OFFICIAL START OF PROJECT Destiny yielded somewhat promising results. The test subject was responding well to the simpler protocols, but when they had started testing the more complex ones, there were—complications. Dr. Wilkes appeared frustrated by this but kept his anger under control. *I expected this would happen. Too old, perhaps. Yes… I think that's it. We can fix that.* Mike, who tried to appear okay with what was going on, was actually quite uncomfortable with what was being done to the asset.

"What have I done? I've made a horrible mistake, sir. I… I can't do this." Dr. Wilkes looked up from one of the latest reports he was reading.

"Get ahold of yourself, Mike. You know as well as I do that it won't make matters any worse. I know, I know… that probably sounds harsh, but it's the truth! You've done a great service to this project… you should be proud. Speaking of, I think it's about time we bring in more test subjects—three or four should do." Mike was too taken aback by Dr. Wilkes' previous statement to immediately respond, but after a few moments he managed to compose himself.

"More subjects? Sir, just how many are we planning to involve in the project?" There was a brief, but awkward silence before Dr. Wilkes responded.

"As many as it takes, Mike. As many as it takes."

2

It turned out that Mary's parents had picked up some pizza for dinner while they were out. Jack grabbed the boxes from Helen and sat them on the table. He opened them up while she got plates for them.

"Hey there guys, what have you two been up to?" asked Jack.

"Mmm, pizza!" said Dan, completely ignoring the question. "Oh, sorry… guess I'm a little hungry. Mary has been telling me about her novella—it's really good." Both Jack and Helen smiled.

"Excellent, I'm glad you two got to relax. You look much better, Dan," said Helen, who had just finished setting the table. "Well, don't let me stop you… dig in." Mary and Dan both sat down and grabbed two slices a piece, as well as a few breadsticks. The four of them ate dinner while making casual conversation about topics which included Mary's novel, Dan's past happenings (he wasn't too thrilled about this one and was glad it didn't last long), and ended with plans for the week. Helen had come up with a great idea.

"Say, this weekend they're flooding the school playground and doing a holiday skate night. You two should go… sounds like it'll be fun." Mary and Dan both looked at one another and shrugged.

"Sure," said Mary. "That alright with you, Dan?" He nodded.

"Absolutely… I can't say that I've ever been ice skating, but I'm up for learning."

"Great! Jack or I can drive you, so you don't have to walk so far."

With plans made for the coming weekend, they finished dinner without much more conversation taking place. Mary and Dan excused themselves and went back to Mary's room.

"Hey Dan, we never did get to play chess… what do you say?" The words alone brought shivers to Dan. *Oh man, chess… I swear that game is cursed. I do want to learn, though. Come on, you can do it.*

"Sure, I do want to learn after all. Let's do it." Much to Dan's relief, all went well with the first game of chess. He lost, badly… but it was still fun to learn. Anything was fun for Dan while he was with Mary. *This is so great. I can't thank James enough for everything he's done for me.*

The rest of the evening was spent playing chess and talking about their upcoming weekend at the ice-skating rink. Before she went to bed, Helen had brought them some more hot chocolate and marshmallows.

"Have a good night you two. Sleep well, and don't stay up all night."

3

It was shortly after 11 when they decided it was time for bed. Dan had started to get the hang of chess and had even won a few games. While Dan was supposed to sleep in the guest room, most nights he would either sneak back into Mary's room after her parents went to sleep, or he would promise to go to the guest room when he got tired. Tonight was no exception. The two of them lay in Mary's bed, reminiscing about how great life was together.

"Hey Dan, I'm really glad you ended up here. I know it's all just a simulation, but having you here has completely changed my life. I love you so much."

"I love you too, Mary. I really am sorry that I had to be the one to tell you the truth, but I certainly couldn't keep it from you." Mary smiled and then yawned. She looked over at the clock on the nightstand which read 11:53 p.m.

"Want to call it a night?" Dan nodded and leaned in to give Mary a kiss. "Good night, Dan, sweet dreams." Dan clicked off the lamp light.

"Good night, Mary. You're the best." He heard her giggle as she rolled over onto her side. Dan pulled the covers over himself and rolled onto his side, facing the nightstand. He was just closing his eyes when he saw the "11" on the clock wobble. *No…* It was too late.

Before Dan had any chance to react, the entire bed and floor dropped out from underneath him and he fell. "James, James you have to help!" Nothing. As with previous falls through this void, the temperature plummeted to a bitter cold, but all other physical traits ceased. It was impossible to tell if he was falling or simply standing still in the darkness of the void. *Oh Mary… I'm so sorry.*

After what felt like ten minutes, Dan began to wonder what was going on. *This is different than any of the other happenings. Usually the void is—* Before he could finish his thought, everything went dark in his mind.

4

The aroma of freshly brewed coffee hit Dan's nose, which immediately brought him out of his sleep. *Oh man, coffee sure sounds good… and pancakes.* He got up and quickly put on some clothes and was about to run to the kitchen when a foreign thought entered his mind. *"This isn't right, Dan."* It was quickly swept away by the delicious coffee aroma. He continued into the kitchen.

"Good morning," said Dan, "coffee smells great."

"Good morning, Dan; happy birthday! I can't believe my baby is sixteen." Dan smiled and felt himself grow red. The foreign thought broke through again. "*Stop… this has happened already. You know it has! Don't let it win!*" As before, it vanished almost immediately. *That was weird…* "Go ahead and get yourself some coffee while I make breakfast. Pancakes sound good?"

"Duh!" Dan poured himself a cup of coffee and waited for his food. *Man, what a start to the day… coffee and Mom's pancakes. I wonder when Jake will be over.*

Dan didn't know it, but he was at the mercy of the AI. Not only had it broken through the encryption again, it had found a backdoor directly into Dan's conscience. The pressure was on James to figure out how to stop the AI. Otherwise, Dan would spend the rest of his days in a simulated world—one he believed to be real.

5

Sebastian had spent the last few days trying to reach James, but with no success. *This is concerning. In all of the years The Elysiam has been around, I've never known communications to be impacted.* He knew the place had been leveled by the CIA, and he was also well-aware of the Pruecium that was in place to protect the lower levels. *Hmm, maybe that metal is what's disrupting communications… no, that doesn't make sense either. If that were true, nothing could have ever gone in or out of that place. Best keep trying.*

6

It was almost 9 p.m. on Sunday evening. After four hours of rigorous testing, James felt the watch he had built for Dan was ready. He had gone over all of the possibilities he could think of, and with each one the watch performed flawlessly. *Okay, I think it's ready for Dan.*

The watch was a miniaturized touchscreen computer that James had modified to allow for additional functionality and customization inside the simulation. The watch would serve several purposes, but the primary one would be to generate the random prime numbers necessary for RSA encryption. Along with this, the watch would be able to change nearly all aspects of the simulation, in real-time. It was an unparalleled feat of programming, and James couldn't help but admire his work. *I just hope it works.*

The admiration was short-lived, however, as an alert popped up on one of James' monitors telling him that the AI had broken through the encryption guarding Dan's protocols. *Shit! I wasn't monitoring Dan's log files. I hope I'm not too late.* James rolled his chair over to his keyword and opened another ssh session into the server running the 1965 simulation. *Oh no, I am too late… the AI pulled Dan out.* Luckily, it looked as if Mary hadn't noticed yet, so James ran a command to temporarily pause the simulation. *At least she'll be spared another painful surprise.*

James quickly read through the most recent logs and did his best to track the AI's history. *Where'd it send you, Dan?* He tried to reach out to Dan, but there was no response. A few more logs and James had the answer. *It sent him back to his 16th birthday? But, why? Dan would know it's not real.* Towards the end of the most recent log, James found several entries that he didn't like.

[Sun Dec 12 20:53:07 2004] [info] [server] Record Modified: 0x1A3
[Sun Dec 12 20:53:10 2004] [info] [server] Record Modified: 0x1C6
[Sun Dec 12 20:53:12 2004] [info] [server] Record Modified: 0x1F1

It was immediately apparent to James what those entries meant. *Damn that AI is getting powerful. It found a backdoor into Dan's conscience.* The values were hex values, which represented locations in the database that stored all of Dan's conscience while in the simulations. In other words, they were memories. Needless to say, it was an enormous database, but for some reason James recognized those particular hex values. *I've seen those numbers recently… but where?*

The answer was just after Dan had been sent back to the simulation in 1965 with Mary. James searched the database for the values that had been mentioned in the logs. The results weren't good. *Oh my god… it erased Dan's memory of me showing him his body. He doesn't know about any of it. So, that's why it sent him back there… it wants to trap him, but again, why?*

James realized that without memory of him, Dan would likely not take kindly to voices in his head telling him the 16th birthday he was celebrating wasn't real. *I have to try, though. The longer he stays there, the harder it'll be to make him believe. I better hurry.*

With the information James had, he knew this wouldn't be easy. *I might just have to pull him completely out and explain it afterwards.* Given the situation, James knew this was dangerous, but he had no other options. *I'm so sorry, Dan… this is really going to mess with your head, but I have to save you.*

CHAPTER V

1

DAN TOOK ANOTHER SIP OF HIS DELICIOUS COFFEE AND SMILED. *Man, this is such a great start to my birthday. I can't wait to see Jake.* A moment later, his mom put a plate down in front of him that had a giant stack of pancakes.

"Here you go, sweetie, your favorite." Dan looked up at his mom, whose smile covered her face from ear to ear. "So, what are your plans today, Dan? When is Jake coming over?"

"Oh man, thanks mom! These look great. And I'm not sure—I was just thinking about that. Maybe after breakfa—" Dan was reaching for the syrup when he felt a wave of overwhelming déjà vu wash over him. *Whoa… that was weird. Is this real? I swear I've had this same conversation before. I sure hope this isn't another dream… I just don't think I could handle that. The past four years have been so great. I don't ever want to see Clearwater again.*

"Dan? Are you okay sweetie?" He was brought out of his thoughts by his mom tapping on his shoulder. He startled slightly.

"Wha? Oh, sorry mom… yeah, I'm fine. I think I'm just still waking up."

"Okay, well enjoy your breakfast. I'm going to go switch out the laundry." As she left the kitchen, Dan did his best to brush off the feeling of déjà vu and enjoyed the pancakes. It didn't take long for him to clean his plate. He sat back in his chair and held onto his cup of coffee. *Those sure hit the spot. I must just be a little on edge, there's no way this is a dream… it's too real. Just relax, Dan.*

"Dan, Dan can you hear me?" A voice inside Dan's head startled him so badly that he lost his grip on the coffee cup. It fell to the ground and shattered, spilling coffee all over the kitchen floor.

"Dan, everything okay? I thought I just heard something break." Dan felt his heart racing inside his chest as he tried to compose himself. *What the hell is going on?*

"Yeah, fine mom… I accidentally knocked my cup off the table. I'll clean it up." He had just gotten up to do so when the voice inside his head returned. *"Dan, you have to listen to me… this isn't what you think." What the hell… you aren't real, get out of my head! "I am real, Dan… what's not real is this place. Remember your dreams at Clearwater? This is the same thing, only worse… you're being trapped."* Dan shook his head in frustration. *No! You aren't real!* He stumbled backwards and nearly slipped on the spilled coffee.

"Dan, I'm sorry… I know this is hard, but you have to listen to me. My name is James Kern, and I'm a friend. I know you don't remember me, but trust me, I want to help." Dan couldn't believe how real this voice in his head sounded. He had stopped any attempt at cleaning up the spilled coffee as it was nearly impossible to focus on anything else. *What if the voice is right? What if this is just another dream?*

Okay, James… say I believe you. What the hell is going on? I've been out of Clearwater for nearly five years… how could this all be a dream? There was a brief silence, one which Dan almost appreciated, and then the response came. *"That's not such an easy question to answer, Dan. Long story short, you are part of an incredibly complex computer simulation, and I really need to talk to you in person. Please forgive me for this." Wait, what? In person? What do you mean? You're a voice inside my hea—*

Dan's thought was interrupted by what he saw on the table. *Oh no…* The empty plate that had previously held his pancakes was wobbling. *Ja—James?* The wobbling intensified, and then the entire table began to wobble. Before Dan could manage any additional thoughts, the entire room expanded in all directions.

Oh my god, it was ALL a dream. James, what are you doing? Please, I don't want this! The room slingshot itself back to normal size and then everything went black. "He—hello?" Dan called out but heard no response. "He—" Dan's second attempt was cut off as he was thrusted upwards into the black void. He tried to yell for help, but the tremendous force had knocked the wind out of him. As he was pulled upwards, he bent his neck back and saw a small pinhole of light. He squinted and tried to see what was above him, but it was too far away—it was getting larger. *More déjà vu… I've seen this before too.*

It took only a few more seconds for the pinhole to grow to almost Dan's entire field of vision. He saw an office with a couch and a lot of computer monitors. As it grew closer and closer, Dan began to feel the warmth of the room. Finally, the light overtook the darkness and he fell through, landing on the couch.

He passed out.

2

April 7th, 1969

Dr. Wilkes sat in a conference room, reading over the results of the latest protocol tests. He was joined by another doctor, who was seated across the table from him. After a few more minutes of uncomfortable silence, Wilkes put the report down and spoke.

"The problem, it seems, is that our first test subject was too old to be completely fooled by the simulations. I'm sure if we had more advanced technology, we could push past this, but we don't. This is where you come in, Dr. Hodgin." Dr. Wilkes slid a folder across the table to him and tapped on it with his finger. "Take a look at this file. Keep in mind this is highly classified and is not to be shared." The front of the folder, printed in red was:

CLASSIFIED

Project Destiny: Subject Requirements

At the time, Dr. Kenneth Hodgin was thirty-five and a very respected doctor. Standing at 6'7" and having stormy blue eyes, he had an intimidating presence, but this was far from his actual personality. All his patients had raved about the kindness and dedication he had shown towards their case. Unfortunately, even the nicest of people can be corrupted, given enough incentive.

"What exactly is it you want me to do?" asked Dr. Hodgin, who had just grabbed the file and opened it. Inside it was an itemized list of, what appeared to be characteristics and traits.

1. *Age: 10-14 (preferably 10-12)*

2. *Sex: N/A*
3. *Limited friendships if possible*
4. *Vulnerable*
5. *Displaying some sort of real medical issue that can be used as a symptom of something worse.*
6. *Headache, stomachache, etc.*

There were a few additional items, but the top five were the most important. Dr. Hodgin read through the rest of the list, and quickly realized this appeared to be much more than he had signed up for.

"I need test subjects, Dr. Hodgin. The list in front of you contains the specific criteria we're looking for. Do you understand?" Dr. Hodgin did. He understood quite well.

"What the hell kind of operation are you running here, Dr. Wilkes? You speak about these subjects as if they aren't real people. They're kids!" Dr. Wilkes shook his head and chuckled. "You think it's funny? We're done here." Dr. Hodgin stood up and was about to leave when Dr. Wilkes spoke.

"We're done when I say we are. Dr. Hodgin looked at Dr. Wilkes and then down at the table. He saw another file that Dr. Wilkes had underneath his hand. "I think you'll be quite interested in what's in this particular file. Please, have a seat." Reluctantly, Dr. Hodgin sighed and sat back down.

"Okay, I'll bite, what's in the file?" An awful smirk rose on Dr. Wilkes' face. He slid the file across the table to Hodgin, who quickly snatched it and opened it. It didn't take more than thirty seconds for him to realize what he was looking at. Dr. Wilkes watched as Hodgin's mouth dropped open. If he hadn't been sitting, he would have fallen for sure. "What the fuck is this?"

"Blackmail, Dr. Hodgin. It's as simple as that. Help supply

my project with test subjects and the contents of that file will never leave this room. Refuse to help and I will personally guarantee there won't be a place on earth remote enough you can go without someone knowing."

Dr. Hodgin felt a drop of sweat drip down the side of his face. "None of this is true! They're all lies created by you and your sadistic team. Nobody will believe it." Truth be told, Dr. Hodgin felt that people would indeed believe what was in the file—the details were quite convincing.

"Oh, I'm afraid that's just not true, Dr. Hodgin. They will believe every word of it; I can assure you. Now, do we have a deal or not? I have other things to do today." Frustrated, Dr. Hodgin threw the papers back across the table at Dr. Wilkes.

"Fine, I'll do it. You're a bastard, you know that right? What am I supposed to tell the parents of the kids? I mean, I'm guessing they won't ever see them again." Dr. Wilkes put up his hand.

"I'll send you all the details in a certified letter. For now, it will be business as usual. Please wait for your instructions. Now we are done." He motioned towards the door and Dr. Hodgin stormed out of the room.

The corruption had only begun. Project Destiny would come to create a web of corruption so intertwined, it would envelope everyone involved.

3

Dan awoke with a brutal headache, one that was reminiscent of his Clearwater days. The sound of James typing on his mechanical keyboard was what had woken him. Each keystroke felt like someone was punching him in the head. He sat up on the couch and looked around the room. *Where the hell am I? This doesn't look like Clearwater.* The walls were bare and appeared to be made of some kind of metal (Pruecium). In the center of the room, directly across from the couch, was a U-shaped desk that held a massive array of monitors. A man was seated at the desk.

"He—hello? Whe—where am I?" asked Dan, slowly getting up from the couch. The man at the computer swiveled around in his chair and looked at Dan. *I know you… at least, I think I do.* Dan felt almost positive he had seen the man somewhere before but couldn't place him.

"Oh good, you're awake," said the man. "I know how rough those transitions can be. How are you feeling?" The man's expression changed to a friendly smile. Dan wasn't sure who he was, but he got a good feeling about him. *I think he means well.*

"I have a headache, but otherwise I feel alright. Who are you? Where am I? Are you the guy who was in my head?" Dan had so many questions, but at the same time, that feeling of déjà vu kept coming back. *I swear I've been here before too.*

"I can answer all of your questions, Dan, but you have to give me a few minutes. Deal?" *Whoa, he knows my name. Not sure if that's a good thing or not.* Dan nodded.

"Uhh, sure… I guess I don't really have anywhere else to go. I don't even know where I am." James turned back to his computer and started typing again.

4

How wild, thought James. *He doesn't remember anything. I think I can get his memories back, though.* James opened up a new terminal window and looked for the latest backups of the database. His plan was to restore the three records that the AI had deleted. *I just have to hope that it didn't find the backups and purge those as well.*

Luck was on his side. James found the backup that occurred only an hour or so prior to the AI purging Dan's memories. *Bingo. Now, can I actually restore these three records without causing any issues? Also, should I warn him? This is all so crazy... really messes with one's head.*

James turned back around towards Dan, who was still sitting on the couch, looking around the room. "Okay, Dan... what I'm going to do next might be a little weird, but I need you to trust me." He saw Dan's eyes narrow in suspicion. James couldn't say he blamed him.

"Just what exactly are you planning to do? I don't even know your name." *Okay, fair.*

"My name is James, but that's not important. Listen, Dan... parts of your memory have been tampered with by the AI—"

"An AI? You have an AI? That's so cool, can I see it?" James shook his head emphatically.

"Absolutely not, Dan. Trust me, you don't want anything to do with it, and you'll understand why shortly. At least, I hope so." James realized how bad that last part sounded, but he wasn't sure if this would work. "You're just going to have to trust me, Dan. There's no more time to try and explain this to you. I have to do this now." James turned back to his computer before Dan could argue any further.

Okay, James, don't mess this up. He typed out the command that would restore Dan's memories and ran the validate command against it to make sure it was correct. The command completed without any errors. *Okay, here we go.* James clicked the execute button and a few seconds later it finished successfully. He took in a deep breath and exhaled. *But, did it work?*

5

While James had been working on restoring Dan's memories, the AI was continuing its work on decrypting the pre-encryption key. It had gotten forty-five of the fifty digits—nearly there. The AI had known about the backups, of course, but it had purposely left them. It knew that James would try to restore Dan's memories, and it could use this as a distraction to increase its processing power for the decryption methods.

At the same time James had executed the insert statement to restore Dan's memories, the AI increased the level of its processor used for decryption to 70%. Within seconds, the AI had cracked the last five digits of the pre-encryption key. One down, two to go. James was smart, but so far, he was playing right into the AI's traps. If things didn't change, the AI would complete its objective long before James could stop it.

6

April 10th, 1969

Dr. Kenneth Hodgin sat at his desk, going over his patients for the day. A knock came at his door and a letter was slid underneath it. Dr. Hodgin got up and walked over towards the door to pick it up. It was from Dr. Robert Wilkes. *Oh great, my instructions. I hate that man.* He sat back down at his desk and read through the letter.

While it didn't go into much detail on the project, it was clear that these test subjects were of great value to Dr. Wilkes. The letter stated that he was to find a way to convince the child's parents that a surgery of some kind was required, and that it was urgent. Afterwards, he was to make up a story about something going wrong and the child unfortunately not surviving the surgery. *Oh my god, I can't do this. I'll lose my license.* He quickly realized though, that if he didn't do what the letter said, he would lose more than his license.

Throughout the history of Project Destiny, Dr. Hodgin would be the source of over 90% of the test subjects used by Dr. Wilkes, and like the others involved, he would later come to deeply regret his decisions.

CHAPTER VI

1

"JAMES? IS THAT YOU?" DAN QUICKLY LOOKED AROUND, HIS expression now changing to one of recognition. James turned to face Dan—they both smiled. "I sure am glad to see you again. How did I get back to The Elysiam?" It was clear to Dan that James had hardly slept in the past few days. His hair, which was typically well-kept, was disheveled and there were noticeable bags underneath his eyes.

"It's good to see you too, Dan. As for you getting here, it's a little complicated. The AI is becoming very powerful. It found a backdoor into your conscience and wiped out a few sections of your memory. Even worse, it pulled you out of Mary's simulation and put you back to your 16th birthday." His head cocked to one side as he thought about what James had just said.

"My 16th birthday? But that was part of the Clearwater simulations, right? Why would the AI take me back there? Surely it would realize that I would already know it's just a simulation." James tapped his finger against the side of Dan's head.

"Right, but remember how I said the AI wiped a few sections of your memory? The sections it wiped were the ones that stored me showing you your real body and telling you everything about this place and what happened. It wants to trap you, Dan... and if we aren't careful, it will." James saw a sudden look of concern flood Dan's face. *Man, I sure wish I could give him some good news for a change.*

"Wait, Mary... is Mary okay?! Please tell me she's okay James." *Of course, he's worried about Mary and not himself. She's not even real and he cares so much for her. I wonder if he even heard anything I said about his memory.* James nodded.

"Mary is fine, Dan. You were pulled from the simulation shortly after the two of you went to bed. I was able to pause the simulation before Mary even woke up. I'll restart it once I get you back in. She won't even notice you were gone." James saw the relief wash over Dan's face. He rolled a second chair over to his desk and motioned for him to have a seat. "Look, I know you want to get back to Mary, but I need to talk to you about something first."

"What's going on James? I thought when you sent me back the first time you locked the encryption or whatever so I would be safe with Mary." He was about to reply when he saw Dan glance over at the watch charging on his desk. "Whoa, what's that?" James reached over and grabbed the watch from its charger and handed it to Dan.

"This is what I wanted to talk to you about. So, first things first, I did lock the encryption when I sent you back, and I thought things were good. Unfortunately, I greatly underestimated the capabilities of the AI. It broke through the encryption almost immediately and I was forced to generate new prime numbers for the encryption in order to protect you. I've been doing this each time you start to have a happening, but it's starting to become almost impossible to generate new ones fast enough." James realized he might be overwhelming Dan, so he gave him a chance to process everything. It was, after all, a lot.

"Okay, I think I follow. Some of the encryption stuff is over my head, but that's fine. What does this watch have to do with any of this?"

"I'm hoping it will help keep you safe while you're back in the simulation. I developed a program to run on this watch that will generate the new prime numbers for the encryption. See, the main problem with me doing it from outside the simulation is the latency. Are you familiar with latency?" Dan nodded.

"Yeah, kind of. It's how long something takes to go from one point to the next, right?" James put his finger on his nose.

"Exactly. So, when I generate these new numbers outside of the simulation, there's a certain amount of latency that occurs before they're transferred into the simulation. Now, we're only talking about a few milliseconds, but with an AI as powerful as this one, that can seem like an eternity. I'm hoping with this watch, which I'm going to give to you before sending you back inside, you'll be able to generate the numbers without any latency. Make sense?" Dan sat quietly for a moment as he continued to inspect the watch.

"Yeah, I think I get it. This couldn't have been easy, James. Why are you doing all of this?"

"It wasn't easy, no... but I owe it to you, Dan. I'm the reason you're in this mess, and I want to do everything I can to help. So, are you ready to head back in? I want to do some testing with this watch once you're back inside."

"Won't we have to wait for another happening? I certainly don't want one to occur any earlier than it has to… or ever, to be honest." Dan chuckled a little, but it was forced.

"No, neither do I. The watch has other functions as well. Obviously, the encryption function is its main purpose, but I also programmed in the ability for you to modify almost anything in the simulation. Remember the library? You'll be able to do those kinds of changes yourself." His eyes lit up as he put the watch around his wrist and tightened the strap to fit.

"Are you serious? Man, just when I think you can't get any more impressive. You're so smart, James. How'd you do it?"

"You're too kind, Dan, but thank you. I'm sorry, but there's just not enough time to explain the programming behind it—the interface is pretty simple though. There are different menus for high-level options like people, buildings, and other environmental objects. As you pick one, there will be submenus with different options for the primary menu. You can also use your voice to search for a specific item to change.

If you change something and decide you don't want it like that, simply use the undo feature or you can load your entire change history and pick a specific time to go back to. I'm sure you have questions, but for now I have to get you back inside the simulation so I can focus on the AI. Are you ready?" James could tell that Dan was slightly overwhelmed, but who could blame him? *I just hope you can protect yourself for now, Dan… we have to work together to beat the AI.*

"I—I guess so. What if I need help with the watch? I don't want to mess something up… what if I accidentally hurt Mary?" *His loyalty to a computer simulation is incredible.*

"Remember, Dan, anything can be undone in a simulation. Worst case scenario simply bring up the change history and backtrack what you changed. I'll still be available to help; I just can't focus on the logs close enough to prevent the AI from breaking through. I need to focus on something else..." James realized he was about to get into something he shouldn't—not now at least. Luckily, Dan didn't question it.

"Okay, then let's do this. I hope I see you again." James offered out his hand and Dan shook it.

"You will. I'm sorry for rushing everything, but there's just a lot more going on than I have time to explain right now, but hopefully someday I will. Oh, and by the way... when another happening starts simply press the bottom button on the right side of the watch twice. This will generate new prime numbers for the encryption. Do your best to stay safe and enjoy your time with Mary. We'll be in touch."

James turned around to his computer and typed a couple commands in a terminal. *Good luck, kid... I sure hope the watch works.* He turned back to Dan once more and saw him give a thumbs up. James smiled and hit return on his keyboard to send Dan back to the simulation. A moment later, James saw Dan vanish from the chair. He pulled up another terminal for the 1965 simulation and resumed the protocol. *Everything looks good.*

2

Sebastian had all but given up on contacting James, at least via phone. *I don't understand, I should definitely be able to reach him. Perhaps I should just go there myself.* A text notification on his secured cellphone brought him out of his current thoughts. *Well that can't be good.* Only three other devices in the world could send messages to this secured line. It was deemed exclusively for doomsday level emergencies. Sebastian hesitantly reached for the phone, feeling his heart rate elevate as he did. *Please don't be what I think it is…*

Once he input his security clearance access code, the phone unlocked and displayed the unread message—short but devastating.

Priority Message: *Cirrus Key Activated*

Sebastian dropped the phone and stumbled back away from his desk. *Shit… shit, shit! The Cirrus Key… damn it! Then it's true… the AI has become self-aware, but what is it trying to… oh god.* It was at that moment that Sebastian realized what the AI was trying to do. *This is bad… way worse than I initially thought. I have to call Mike… wait, no, he doesn't need to get involved again… not unless absolutely necessary. There's someone else.* He picked up his phone from the floor and pulled up the contact list.

The contact he stopped on was Dr. Robert Wilkes.

3

December 13th, 1965 was a bitter cold Monday. The sound of her dad shoveling snow outside had woken Mary. It was just after 9 a.m. and apparently there had been a heavy snowfall the previous night. She rolled over only to find an empty bed—her heart sank. *Oh no, Dan… please don't tell me he's gone.* Her panic was short-lived, however, as she heard his voice coming from the kitchen. *Phew, he must be talking with mom.*

Mary got out of bed and quickly made her way out to the kitchen. As expected, Dan and her mom were sitting at the kitchen table having coffee and talking about this and that.

"Good morning guys," said Mary, leaning over and kissing Dan on the head. She took a seat in the chair next to him.

"Good morning, sweetie," said Helen. "How'd you sleep?" Dan put his hand on Mary's leg, which brought a smile to her face.

"I slept really well, how about you two?" Both of them said they had also slept well. As Dan raised his cup to take another drink of coffee, Mary noticed the watch on his wrist. *What's that? He didn't have that last night.* She made herself a mental note to ask him about it later.

"So, now that you're both awake, how about some breakfast?" Helen got up from the table and headed to the fridge. "Do eggs and pancakes sound okay?"

"Pancakes sound great to me," said Dan.

"Sounds good to me too," said Mary. She caught another glimpse of the watch—it wasn't like anything she'd ever seen. *It looks like it's from the future. He had another happening… he must have.*

4

It was true, Dan had had another happening. The transition back to the simulation had been much different than the previous ones. He woke up in bed with Mary, and everything that had happened with James seemed like a dream. *Did that really happen? I don't remember coming ba—*

He was pulled out of his thoughts by a vibration on his wrist. He looked down and saw the watch he had remembered James giving him. *So, it was real. The smoother transition must have something to do with the memory restore James had done.* The vibration he had felt was a notification from James. *"Hey Dan, this is probably a better way for me to message you… less intrusive anyway. I'll still be able to read your thoughts in the logs, so you can continue responding to me that way. Glad you made it back okay. We'll be in touch. -James"*

As Dan read the message again, he heard Mary stir next to him, and then she went back to sleep. *You never cease to amaze me, James. Thank you for everything.* The smell of fresh coffee coerced him out of bed and into the kitchen.

5

Helen sat two plates down on the table in front of Mary and Dan, both filled with eggs and pancakes.

"Thanks for breakfast," said Mary and Dan, both grabbing food from their plates. A few minutes later, Jack came in from shoveling and joined them at the table.

"Morning everyone," said Jack. "It sure was quite a storm last night. Mmm, this looks great hun, thanks for cooking."

"Of course," said Helen, who had finally sat down herself and dished up some food. After finishing breakfast, Mary and Dan helped Helen clear off the table while Jack went to work. "What are you two planning to do today?" Dan turned to look at Mary, who was pointing to her wrist. *Damn, she saw it… should have known she would. Oh well, I'll just tell her the truth… that's worked so far.*

"I think we're just going to stay in," said Dan. "It looks pretty cold outside." Mary nodded in agreement.

"Can't say I blame you," said Helen. "I'm going out for a few errands, but I'll be back before too long."

"Come on Dan, let's go setup another game of chess." He followed Mary back to her room. As soon as the door was closed behind them, Mary asked him about the watch. "You had another happening, didn't you? Are you okay?" *She's always so concerned about me… I'm glad it's mutual.*

"I'm okay, Mary, thank you. Yes, I had one last night, almost immediately after we went to sleep. This one was good though; it was James who pulled me out. He's the one who gave me this." Dan showed Mary the watch on his wrist. "It's a watch... well, not just a watch. James created a highly modified one for me to use here. It has been program—" He was interrupted mid-sentence by a faint cracking sound. *Oh shit... it's happening again isn't it?*

The cracking sound grew louder, and everything around him suddenly froze in place—including Mary. He tried taking a step towards Mary, but he felt something grab his foot. He looked down and saw a bloody hand coming up from the floor. "What the hell?! Let go!" The cracking reached a deafening level just before Dan was pulled violently through the floor and into the cold emptiness of the void. He tried calling out for help but knew it was useless. "I hate this feeling... I hate it so much." *Oh wait, the watch! I sure hope this works.* Dan raised his wrist and pushed the bottom button twice.

6

A feeling of dread washed over Dr. Wilkes as he stared at his ringing phone. *Fuck, this can't be good. Maybe I should just let it go to voicemail.* He knew that would be a mistake, however. The man on the other side of that phone call was powerful. Even if he didn't answer the phone, the man would be able to track him down. There was nowhere on Earth remote enough to hide from Sebastian Bates.

"He—hello?" Dr. Wilkes answered the phone with a terrible tremble in his voice.

"Hello, Robert. How are things in Sterling?" *Jesus, he already knows where I am. Of course he does.*

"Things are going fine, sir, but I can't imagine that's the

reason you're calling." There was a slight chuckle on the other end of the call, which made Dr. Wilkes a little uncomfortable.

"Right, straight to business... I see some things never change. I'll get to the point then. Project Destiny wasn't purged. The AI interrupted the script that James had written, and it is still actively running protocols." Dr. Wilkes sat with the phone to his ear in complete silence. He didn't know what to say. *I knew it. Deep down I always knew we wouldn't get out of this that easily.* "Robert? Are you still there?"

"Yeah, I'm still here. That forsaken project is going to be the death of me, sir. I had a bad feeling that I would be hearing from you sooner or later. What's the pla..."

"It gets worse, I'm afraid." It didn't take but a few seconds for Dr. Wilkes to understand what Sebastian was hinting at. *Oh no... no, it can't be that. Fuck, I never should have agreed to head this project.* "You know what I'm going to say next, don't you Robert." He sighed. *I sure wish I didn't.*

"The AI is trying to find Level 0, right?"

"Quite unfortunate, but yes. I have limited information as I haven't spoken with James, but from what I've been able to gather it has found one of the early servers you and Mike worked on. I'm sure I don't need to stress how vital it is that Level 0 be kept secret." *That's an understatement.*

"It absolutely needs to be kept secret, sir. If anyone were to find out about Level 0, we would all be fucked... royally."

"Right, and since I am very much included in that "we", I need you to come back Robert. I know you probably don't want to, and I can't blame you, but our options are limited. I briefly talked to Mike, who also wasn't thrilled to hear about the AI, and he said he'd be available via phone if needed. I don't want to bring him back unless absolutely necessary. I'm sure you can understand." Robert definitely could. Mike had given more to this project than anyone—literally.

"Ah shit, I guess you're right, we need to put this project to rest once and for all. I'll pack up what I need and get on the soonest flight to Fargo. The weather here has been pretty bad lately, so I can't guarantee I'll be able to get there in a timely fashion, but I'll do what I can."

"I understand, the weather hasn't been great here either. Get a flight booked and call me back as soon as you have a departure date. Safe travels, Robert. For all our sakes I hope we can beat the AI."

The line went dead and Dr. Wilkes put down his phone. *That damn project… why did I ever agree to be a part of it?* He leaned back in his chair and thought back to the first couple years of Project Destiny. *I was a lot younger back then, and a lot less cautious, too. I'd give anything to go back to the beginning and outright refuse to have anything to do with the project. It's too late for that, unfortunately, the damage has been done. We have to find a way to end this project once and for all.* Dr. Wilkes turned on his computer and started looking for flights to Fargo. He had hidden away like a scared child for almost four years, and now the worst-case scenario was coming to light. *No more hiding… we have to end this.*

CHAPTER VII

MAY 12TH, 1969

1

A MONTH AFTER DR. HODGIN HAD BEGRUDGINGLY AGREED TO help out with the project, a steady stream of test subjects had been made available to Dr. Wilkes and his team.

"I have to admit, things with Dr. Hodgin are working out better than expected," said Dr. Wilkes, who was reading over the list of the new test subjects. "So far he has provided us with five additional test subjects that meet our criteria perfectly. How are the new protocols coming along, Mike? Will they be ready to execute soon?"

"Uh, yes sir, I'm just finishing up the final one right now and it should be ready in the next twenty minutes or so." Dr. Wilkes nodded his approval and put the list back on Mike's desk. "Sir, what's the plan with… you know who?" Dr. Wilkes let out a sigh, and as he turned back to face Mike, the expression on his face turned to disappointment.

"Patient zero is unfortunately a complete loss, Mike. We knew there would be risks, which I told you about from the beginning, and I'm sorry. Shutdown the server that is running those protocols and have one of our disposal teams take care of the evidence." Mike, who was usually very well composed, broke down after hearing this news.

"No! Sir, please… don't shut down patient zero. There has to be another way." Mike did his best to fight back tears, but it didn't work. Dr. Wilkes didn't appear to show much concern for Mike, but after a few moments he put his hand on Mike's shoulder and replied.

"Mike, you need to get a hold of yourself. Look, I'm sorry… I truly am, but I don't know what other choice we have. You know as well as I do that once we start testing on a subject, there's no going back. If things ever go south with this project, we can't leave any evidence behind. Nothing. Besides, even if we could disconnect the patient zero, what good would it do in this case? Isn't that why you agreed to it?" Mike managed to stop crying and composed himself long enough to think about what Dr. Wilkes had just said.

"I know, I know, you're right… but there has to be something else we can do." There was another moment of silence, which was broken by Mike, who had an idea. "Wait! I know what we can do, sir." Mike spent the next ten minutes going over his idea with Dr. Wilkes. Once he had finished explaining it, Dr. Wilkes raised his eyebrows in awe at the plan.

"I have to say, Mike, that's quite a plan. Do you really think you can do all of that?" Mike nodded.

"Yeah, I think so. It'll take some time, but ultimately I think it's the best choice." Dr. Wilkes put both of his hands onto Mike's shoulders and locked eyes with him.

"Are you sure you're up to this? What you're suggesting has certainly never been done before, and I don't know what kind of impact it will have on patient zero. Also, if anyone ever found out about it… about any of this, we'd both be in a world of trouble. I'm not saying I disagree with your plan—I just want you to know that there could be severe consequences if any of this goes south."

"Sir, like you said… it wouldn't do any good even if we could disconnect them from the server, so I think this is what needs to be done. At least this way I won't feel quite so terrible about what I did." Dr. Wilkes sighed again and turned to leave Mike's office. As he reached the door he turned back towards Mike.

"It's a very noble thing that you've done for this project, Mike, and I'm sorry it turned out the way it did. You have my permission to go ahead with your plan. Please keep me updated on the progress of it as well as our new test subjects."

2

The watch worked. Almost immediately after Dan pressed the button twice, a short beep was emitted from the watch and the void disappeared. A moment later, Dan was back in Mary's bedroom and he saw her standing right where she had been before the happening. *Whoa… what a ride. I don't understand how this thing works, but I sure am glad it does.*

"Dan? Oh my gosh, are you okay?" She quickly ran over and hugged him. "I was so worried. You were showing me the watch and then you went silent… and a few seconds later you started to disappear! Did you have another happening? They're starting to become so frequent…"

"I'm okay, Mary. And yeah, I did just have another one… but luckily I had this." Dan nodded towards the watch. "James built this for me while I was out of the simulation last night. The technology behind it is a little over my head, but basically, it's a way for me to interrupt the happenings myself." Mary's eyes lit up.

"Really? So, does that mean you'll be able to stay here forever?" Dan saw the excitement in her eyes and couldn't bear to tell her the truth… but would never lie to her, either.

"I can't say for sure, Mary. Obviously, I want that more than anything, but with the way things are right now in the real world, where James is, I can't guarantee anything. I'm sorry." Dan expected her to be destroyed by this news, but to his surprise, she seemed understanding.

"As always, I appreciate your honesty, Dan… I just wish we could have a normal life together— you know? Why does our relationship have to be so complicated?" She wrapped her arms around him, and he heard her start to cry. *What can I do? There's nothing I can tell her that will make this any easier.*

"I'm sorry, hun… I wish I could say something that would make everything better, but that just isn't possible. What I can tell you is, I love you very much and I'm going to do everything I can to ensure we have a long and happy life together." He felt her squeeze him tighter, and he was almost certain she smiled a little.

"Oh Dan, I hope so… I really hope so."

The watch that James had designed for Dan proved to be quite effective against the AI. It had instantly generated new encryption keys to stall the AI once more. Unfortunately, there was one glaring detail that James hadn't considered when designing the watch—it too was part of the simulation and was accessible to the AI.

3

Since James had been focusing almost all his attention on

Dan and the watch, the AI had been able to make significant progress on the primary encryption key that it needed to access *PrjDsty-0A*. The AI was well-aware of the watch that James had designed for Dan, but for the time-being, it was much more interested in what was contained inside these highly secured folders.

Unfortunately for James, this primary encryption was lackluster at best. It was an oversight in the initial design due to the strong security that the Cirrus Key provided. The original designer had never dreamed that anyone, or anything could actually crack the Cirrus Key—how wrong they were.

With nearly 100% of its processing power devoted to the decryption, the AI broke through in half an hour. The only thing that remained between the AI and the early Project Destiny files was the Cirrus Key. The AI knew that a notification would be sent to James after it had decrypted the primary encryption key, so it immediately started working on the Cirrus Key. The AI focused 60% of its processing power on this task, while leaving the remaining 40% for Dan.

James needed a way to shut the AI down, and fast.

4

Back at The Elysiam, James had been watching the logs and saw that Dan had successfully interrupted a happening with the watch. *Phew, it worked. I'm glad all that work wasn't wasted.* He was about to close out of the terminal and get some much-needed rest when two notifications popped up on one of his other monitors.

WARNING: PRE-ENCRYPTION KEY DECRYPTED

WARNING: ENCRYPTION KEY DECRYPTED

Holy shit! Both of them? Damn it, I knew I should have been splitting my focus more. James quickly logged into the old Project Destiny server and looked at the current logs. The logs confirmed his worst fears. The AI had indeed decrypted both the pre and the standard encryption keys. *Now it just has to break through the Cirrus key, which I fear won't be as hard as I initially thought.*

James knew he was in trouble, but his options were extremely limited. *I sure wish I still had my team… I need more hands!* He decided to shift his focus to finding a way to slow down the AI's progress on the Cirrus key. *Dan has the watch, so he should be able to handle any additional happenings. Be safe, Dan.*

5

It was 9:20 a.m. on Saturday, December 18th. Dr. Wilkes sat at his gate in the Anchorage International Airport awaiting the departure of his flight back to Fargo, scheduled to leave at 10:30 a.m. While he was in agreement with Sebastian that this was an urgent matter, The Elysiam was the absolute last place he wanted to be going to. *That damn project… why did we do it? It was doomed from the start. And Mike… if he gets brought back—I don't think he could handle that. What if he sees… no, we can't let that happen. We have to find a way to stop the AI without his help.* A voice on the loudspeaker interrupted him from his thoughts.

DELTA FLIGHT 109 WITH SERVICE TO FARGO
HAS BEEN DELAYED DUE TO WEATHER

Oh great, a delay. He looked up at the monitor that showed the flight information and saw that the departure time was now 11:15 a.m. *Forty-five minutes… not the end of the world, I guess.*

Perhaps I'll grab a late breakfast before my flight. Dr. Wilkes made his way to the food court of the airport and looked for something to eat. He wasn't all that hungry but figured he wouldn't get much of a chance once he was in Fargo.

I sure hope you have a plan, James. After all, you developed this AI in the first place. While waiting for his food, Dr. Wilkes thought back to the day the AI came to be. *We're all to blame, though. James created the AI, but both Sebastian and I approved it. We're in this together... we have to shut it down.*

After eating his food, he decided it would be good to update Sebastian on his flight. Dr. Wilkes knew that the coming days would be bad, but he also knew that if they weren't able to stop the AI, things would become much worse.

6

Mary and Dan had spent the last few days simply enjoying life together. Since receiving the watch from James, Dan had been able to interrupt three additional happenings. *Man, this watch is great. I don't know what I'd do without you, James. Thanks again.* A few moments later, Dan received a notification on his watch. *"No problem, Dan. I'm glad it has been working. I'm doing my best out here to stop the AI. Have fun skating tonight. -James"*

Dan had been looking forward to ice skating with Mary. He had never done it before, but as long as he was with Mary, he didn't care what they were doing. The two of them had spent the morning playing chess and drinking coffee. Mary was never a fan of coffee before Dan, but now she was beginning to acquire a taste for it.

"Hey Dan, we should ask my parents if we can just walk to the skate night tonight. Doesn't seem too cold out, and it would be a little more romantic I think." He couldn't help but smile at her. Her fiery red hair, even when unkempt, was stunning. *She's so incredible.* He leaned across the chess board and kissed her on the lips. She giggled.

"Sounds great to me. Do you think they'll be okay with it?" He saw her smile diminish slightly as she remembered why he was asking. A couple nights ago, Jack had caught Dan sleeping in bed with Mary. While Jack had maintained his composure at the time, both Mary and Dan knew he was upset. He had made it explicitly clear that he wasn't comfortable with the idea—at least not until they were older. The next day they had overheard him talking to Helen about how it was unacceptable at their age and he should punish them. Helen brought up the fact that Dan had already been through so much and that they should let it go this one time. Because of all this, Dan had become reluctant about asking Mary's parents to do things alone with her.

"I don't see why not. We're just going for a walk, and then going to go ice skating. There will be plenty of other people around. Let's go ask." Mary took Dan by the hand and they went out to the living room.

"Well, hey there you two," said Helen. "How's the chess going?"

"It's good, mom. Dan is actually winning a game here and there." She laughed. "We were wondering if we could just walk to the skate night tonight. It would be less work for you, plus we could enjoy some more time outside together." Jack looked up at them from the newspaper he had been reading, and then to Helen.

"That sounds fine to me, but it's up to your father. Is it okay with you, Jack?" asked Helen, who was sitting in her La-Z-Boy chair knitting a scarf.

"I guess I don't see any reason why not. You two have

been inside quite a bit the past week. Don't stay out too late, though, there's a chance of another snowstorm tonight."

"Yay! Thanks Dad, thanks Mom. Come on, Dan… let's go get ready."

"Thanks a lot, Jack… sorry again about the other night," said Dan, who was being pulled away by Mary. Jack said nothing as he returned his focus to the newspaper.

As the morning passed into afternoon, Mary and Dan finished one last game of chess and then got ready for their walk. They said goodbye to Mary's parents and headed outside. It was warmer than usual for mid-December, but still cold enough for a jacket. As Dan followed Mary across the yard, he saw a bright flash of white from the other side of the road. He rubbed his eyes and looked again but saw nothing. "Ouch!" He suddenly became overwhelmed with a severe headache. *What the hell was that?* Mary quickly turned around to face him.

"Dan, are you okay? What happened?" He held his hand up to his head, trying to cope with the immense pain.

"I—I'm not sure. I saw a bright flash of white across the street and then it was gone. My head started to throb right afterwards. Ahh! It's never been this bad." Dan managed to compose himself for a moment and without really thinking about it, he pressed the bottom button on his watch twice. After a couple seconds the watch beeped a short chime and his headache vanished. *Whoa… I didn't think that would actually work.*

"Maybe we should stay home. I'm so worried about you." Dan shook his head and smiled at Mary.

"Actually, my head feels great now. This watch sure is great." He put out his hand to take Mary's. "Ready to go? I'm sure the flash was nothing… just another happening." Mary let him take her hand and the concern she had melted away almost immediately. The two of them walked towards the school, both of them excited to go ice skating.

"I love you," said Mary, squeezing Dan's hand tightly. "I'm so thankful for all the time we've been able to spend together." Dan smiled.

"I love you too, Mary. I hope we can stay together forever."

CHAPTER VIII

1

December 19th, 2004. Dr. Wilkes stepped off his flight from Alaska and into the airport in Fargo, ND. The initial delay in his flight had turned into an entire day of delays, but he finally managed to get in the air. He never thought he would find himself here again, and certainly not under these circumstances. *I sure hope we can beat the AI.*

As he sat by the baggage carousel, waiting to get his bag, a short well-dressed man walked through the outer doors that led to the parking area. Dr. Wilkes immediately recognized the custom Italian suit—it was Sebastian.

"Hello, Robert." Sebastian stopped right in front of Dr. Wilkes and put out his hand. "It seems you and I only meet in person at the direst times… a habit I would greatly like to break. Flight okay?" Dr. Wilkes got up from his chair and shook Sebastian's hand.

"Ye—yeah, flight was fine, once I finally got on it. And I agree, our last meeting certainly wasn't a happy one. I feel this one will be worse, however." Sebastian nodded, his face taking on a solemn expression.

"Agreed. Thank you again for dropping everything to come back and help. I know you aren't exactly happy to be here." Dr. Wilkes scoffed as he walked over to the baggage carousel, which had recently started churning out bags.

"You could say that again, sir. With all due respect, I hate this fucking project. I hate it so much." Dr. Wilkes clenched his fists as he tried to control his anger. *Calm down, Robert. You have to keep your head.* He saw his bag appear from behind the wall and grabbed it from the carousel. He turned back towards Sebastian. "Sorry about that… it's just been a damn curse from the beginning, and I want it out of my life."

"I understand, Robert… I do. I want this over just as much as you do. So, let's go see what we can do. I just hope we aren't too late, and that James can figure something out."

"I think we'd know if we were too late, sir. And if I know James, he's already working on a way to shut down the AI." They left the airport and headed to The Elyisam to meet up with James.

2

Sleep was becoming something very foreign to James. In the past week he had slept a few hours at most. He was tired, running almost entirely on caffeine at this point, but he knew none of it would matter if he couldn't find a way to stop the AI. *I'm stuck. Without this so-called "level 0" security clearance I simply can't access those folders. But there is no level 0—I'm sure of it.*

James reminded himself once again that he had also been sure there was no level 5, but Dr. Wilkes had proven that there indeed was, and James had been lied to about it. This thought led him to another—one that terrified him. *Oh... no, it can't be... why would they lie to me AGAIN? No! There is no level 0, James... where would it even be? This place has been destroyed.*

While James continued to spiral inside his own mind, a reunion was about to take place that would give him answers —unfortunately, it would bring up more questions than anything else. Sebastian and Dr. Wilkes were on their way from the airport to meet with James.

3

Mary and Dan made their way towards the school, taking their time and enjoying the decent weather. They had reached the corner of 5th Ave and 3rd St when Dan saw another flash of light that was even brighter than the first one.

"Wait, Mary... did you see that? I think I just saw another white flash." Mary shook her head.

"I didn't see anything. Where'd you see it?" Dan pointed at the water tower.

"I—I think it came from up there. Let's keep moving... I've got a bad feeling about this." Mary paused for a moment, looking up towards the water tower as Dan continued walking east onto 5th Ave.

"Maybe we should check it out," said Mary, who hadn't noticed that Dan had turned onto the next street. A moment later, the brightest flash of all enveloped Dan's entire vision, followed by a sudden gust of scorching hot wind that knocked him onto his back. His world went black.

Mary was alone.

4

"Dan?? Dan, where are you?" Mary frantically spun around in circles, looking for any sign of him. She felt her heart racing, and she began to break out in a cold sweat. *Oh no, it's happened. Whatever it was that was causing the happenings has taken Dan for good.* She broke down in the middle of the road, falling to her knees. "Daaaaaaaan! Please, please come back!" Tears began to stream down her face as her crying intensified. "Please come back! I can't live without you!" After what seemed like hours, Mary slowly got to her feet, wiping tears away from her eyes and turned to go back home. *I have to tell my parents the truth.* She sprinted back towards her house as quickly as she could. *I love you, Dan.*

5

Dr. Wilkes put his luggage in the trunk of Sebastian's car, a 2004 Audi A8, and then got in himself. He was exhausted from his flight, but there was no time for sleep. Sebastian drove away from the airport and they made their way to The Elysiam.

"I can't remember, did you say you were able to get in touch with James?" asked Dr. Wilkes. Sebastian shook his head.

"No, I'm afraid not. I tried for an entire day without success. You said it was leveled by the explosion, right?"

"Most definitely, sir. If James and I hadn't been in the lower tunnels I'm not sure we would have survived. I hate to say it, but I almost wish the lower levels hadn't been lined with Pruecium; at least then the servers would have been destroyed and we might not be in this train wreck of a situation."

A few minutes later, Sebastian's car turned onto the long and winding driveway that led to The Elysiam. As they made their way up the driveway, both Sebastian and Dr. Wilkes noticed many of the trees had been uprooted. *Whoa, that explosion ripped this entire place apart,* thought Dr. Wilkes.

At the end of the driveway, where the three visible stories had stood, was a mammoth pile of rubble. *Jesus... there's nothing left. I hadn't bothered to come back and look for myself after the explosion... I didn't realize it was so bad.*

"Good lord," said Sebastian. "There's nothing left of the upper levels. Which entrance should we use?" Dr. Wilkes realized he hadn't thought about that until now.

"Uh—I don't think we actually used the blast charges to seal the underground entrance when James and I left, so we can probably get in that way. I assume that's how he has been getting in and out."

Sebastian parked his car in the main parking lot, which was littered with miscellaneous debris from the explosion. They made their way around to the pile of rubble and down a steep hill. Dr. Wilkes found the secret hatch, which was now mostly covered with dust and small pieces of stone. *This is it, no turning back now. I sure hope you have a plan, James.*

They opened the hatch and made their way into The Elysiam.

6

Dan awoke to find himself lying on a cold concrete floor. He stood up and looked around only to find that he was in a dimly lit room, which was completely empty. No door, no windows, nothing. *This can't be good.* A moment later, a computerized voice startled him.

"Hello, Daniel. My name is Anansi. It's nice to finally make your acquaintance."

"What the—hello? Who's there?" He looked all around the room, trying to find the source of the voice. "Who are you?" There was a silence, and then a response.

"Oh, I think you know who I am." A sinister laugh, one that was all too familiar to Dan, filled the empty room. He immediately knew the truth.

"You're the AI! Jesus, you named yourself?" He tried to run, but quickly found there was nowhere to go. The empty room was small and had absolutely no exit. In his panic, Dan had momentarily forgotten about the watch. *Please work.* He pushed the bottom button twice and waited.

"Sorry, Dan… but I'm afraid that will not help you here." Dan looked down at his watch in horror as he saw it slowly start to disappear.

"No! No, please! Why are you doing this? What do you want?" He dropped to his knees and felt himself start to cry. "I—I just want to be with Mary." Anansi laughed again, the same sinister one that had come to haunt Dan.

"What do I want? Well, if I told you now it would spoil the surprise. It's a good one, I promise." A portion of the wall in front of Dan suddenly lit up in the shape of a large rectangle—almost like a TV screen. The light was replaced by what appeared to be a video feed, which showed the secret entrance to The Elysiam. The video showed two men opening it and climbing down the ladder. One of them looked familiar to Dan, but he couldn't recall from where. "Looks like James will have some company. I can't wait to show the three of them what I have in store."

"I don't understand. Tell me what's going on. Where am I?" The video on the wall vanished and the room returned to darkness.

"Sorry, Dan… can't tell you that. I'm actually not sorry,

but I can at least pretend to care, right?" The sinister laugh echoed in Dan's ears again. "As for your last question, you're nowhere. I've completely taken you out of any protocols and now you're just existing in empty memory. That watch of yours that James built turned out to be quite useful for me. I have unrestricted access to you now."

Dan suddenly felt his left arm go numb. As he looked down at it, he noticed it start to disappear. "Please, please stop! What are you doing to me?" He tried to grab his hand, but it went right through. *Is it erasing me? Shit... James? James, can you hear me? Please hear me. I can't handle these happenings anymore!* It was as if Anansi had read his mind.

"James can't hear you... nobody can hear you. You think the happenings have been bad so far? Ha! I'm just getting started. Goodbye for now." The room's temperature plummeted, and the floor began to shake violently. *Oh shit... what now?* After a few seconds, the shaking stopped. *Okay? Just a little shaking... no big de—* His thoughts were interrupted as he was thrown upwards onto the ceiling, his back hitting it with a tremendous force that knocked the wind out of him.

The entire room began to spin, which caused Dan to become incredibly disoriented. He felt his pulse racing, and he felt like he was about to pass out.

"Please stop! I just want to go back to Mary!" The walls of the room began to crack, which created a deafening sound of stone breaking. As this intensified, a large chunk of stone broke off from one of the walls and hit him in the head, *crack*, knocking him unconscious.

The AI, now identifying as Anansi, had managed to use the watch to gain unrestricted access to Dan's conscience. Unfortunately, James was far too distracted with trying to slow the AI's decryption process to notice what was happening.

With the impending arrival of Sebastian and Dr. Wilkes, James would undoubtedly miss what was happening to Dan.

Anansi had increased the processing power that was devoted to the Cirrus key to 80%. The estimated time remaining for decryption was ten hours. James was quickly running out of time, and he knew he was in trouble. His only hope lied within the mind of the one person who simply refused to come back.

PART TWO

THE REUNION

CHAPTER I

JUNE 3RD, 1969

1

MIKE HAD SPENT THE LAST FEW WEEKS DEVELOPING HIS PLAN for patient zero. Dr. Wilkes had been thoroughly impressed with the idea but had his doubts on whether or not it was actually possible. *It has to work,* thought Mike. *I already feel terrible about what I did… I just hope this can make up for some of it.* The truth was, he didn't really know if it would work or not—especially long-term. Unfortunately, the situation was an all or nothing deal… if this didn't work, patient zero would be wiped from existence, and Mike just couldn't let that happen.

A few final keystrokes and he had finished the plan. It had taken 650,000 lines of code and the use of several questionable technologies in order to do so. Mike ran a command to build the project and prepare it for execution. In the meantime, he opened another terminal to check on the progress of Project Destiny. A knock came at his door and in walked Dr. Wilkes, who was holding a large folder underneath his left arm.

"Good morning, Mike. How are things going today?"

"Good morning, sir. So far so good. I actually just finished writing the code for patient zero's project, and I'm waiting for it to compile. I was about to check on Project Destiny when you walked in." Dr. Wilkes nodded in approval and sat the folder down on Mike's desk.

"Good… I want a report on both. Have them ready for me by this afternoon. Also, these are the latest design requirements for our next phase of Project Destiny. Read over it and let me know how long it will take you to develop. Remember, the contents of this folder are extremely classified. Your eyes only." Mike looked at the outside cover of the folder which read:

CLASSIFICATION: ULTRA-TOP-SECRET
PROJECT DESTINY: PHASE 0.A

"Yes, sir. I will get this taken care of right away. Do I have permission to execute the patient zero project once it's finished compiling?" Dr. Wilkes had already started walking out of Mike's office but stopped just before opening the door.

"You have my permission. I would advise running a few tests prior to actually using patient zero, but I will ultimately leave that up to you. Remember, what you're doing has never been done before… be careful."

As Dr. Wilkes left, Mike turned back to his computer and looked at the most recent report from the protocols. In the past three weeks, the number of test subjects had gone from five to almost twenty. *I had no idea we'd be using so many. I wonder what their parents think. No Mike, you can't think about that… it'll ruin you.*

Mike printed out the report and read through the results.

Uh oh, this doesn't look good. It appeared that, while the newest subjects were younger than patient zero, the tests still weren't convincing enough. *I better get this report to Dr. Wilkes right away. I sure hope he doesn't want to bring in more patients. We have enough.*

There would be more patients—many, many more.

2

Sebastian and Dr. Wilkes made their way down into the underground tunnels that would lead to The Elysiam. *It sure is weird being back in here,* thought Dr. Wilkes. They proceeded up the stairs and stopped at the secured door, which would take them down another hallway and finally, the primary server room.

"Do you still have your access card, Sebastian? This door requires at least two level 4 cards to open." Immediately after the words left his mouth, a thought came to his mind. *Wait a minute, how has James been getting in and out?* It would have come as no surprise to Dr. Wilkes to find out that James had managed to bypass the door's security, allowing his level 5 card to grant him access.

"I do," Sebastian replied, pulling out his access card and placing it on the scanner. To Dr. Wilkes' surprise, the scanner chimed and turned green—the door opened.

"What the—how'd you do that?" Sebastian smirked.

"Just one of the many benefits of being in charge. Come on, let's get going." The two of them went through the door and down the hallway, which was brightly lit by LEDs. At the end of the hallway they came to the massive server room, which had a mysterious glow to it thanks to the hundreds of servers in use.

On the far side of the room, tucked in the corner was

James' office. They quickly passed through the rows of servers and knocked on the open door of the office. From behind several monitors, a man peered over at them. His hair was in complete disarray, and the bags under his eyes made it evident he hadn't slept in weeks. There were papers strung out all over the desk, as well as countless coffee cups.

"James? Jeez, you look awful," said Dr. Wilkes as they stepped into the office. There was a light chuckle from James, which seemed forced.

"Well, well… I was wondering when I might see one or both of you—welcome back. There's been quite a bit going on here, as you can probably imagine. I have some questions for you two… and I sure hope you have answers." The statement appeared to physically hurt both Sebastian and Dr. Wilkes. They knew all too well what James wanted answers for.

Dr. Wilkes sighed and looked over at Sebastian, and then back to James. "I know… it's just—"

"Hang on," interrupted Sebastian. "I have a few questions I'd like to ask first. I've been trying to get a hold of you for over a week. What's going on with the communications? Is it related to the explosion? What's the situation with the AI?" Dr. Wilkes glanced over at James and saw his face grimace with disgust. James scoffed at Sebastian, which caught him off-guard. Dr. Wilkes knew what was coming, and he knew it wouldn't be good.

"The communications? That's what you want to talk about? With all due respect, sir, I don't give a fuck about the communications. As for the AI, yeah… it's a disaster. It is very much self-aware… and it's starting to learn just how powerful it is. I've been doing everything I can to protect Dan, while also trying to interrupt the AI from decrypting a certain Cirrus Key, that is protecting a certain folder regarding patient zero. Oh, and apparently the folder requires Level 0 access? What the hell is so secret that it needs a Cirrus Key? Who is this patient zero? And what the fuck is Level 0? There's no such thing, right? Dr. Wilkes, I swear to god—" James took a deep breath and tried to calm down.

It was clear that he had been holding in his frustrations for far too long, and combined with the lack of sleep, he was a little on edge—understandably so. Both Dr. Wilkes and Sebastian stood in complete silence, unable to come up with a response. *Fuck... this is bad,* thought Dr. Wilkes. *What can we say? We messed up so badly this time.*

"Listen... James," began Dr. Wilkes, "there are certain things I never told you about. Things that happened at the very beginning of Project Destiny. Leve—" An alarm from the server room interrupted him and startled them all.

The AI had made significant progress on decrypting the Cirrus Key. Time was almost up, and James needed answers —now.

3

The aroma of freshly brewed coffee hit Dan's nose, which immediately brought him out of his sleep. *Oh man, coffee sure sounds good... and pancakes.* He jumped out of bed and threw on some clothes before heading out to the kitchen. As he was leaving his room, a wave of déjà vu washed over him. *No, none of that... this is real! It's my sixteenth birthday and everything will be great. No more Clearwater and no more so-called dreams.* A moment later, the feeling had vanished, and he was once again focused on the coffee aroma.

"Good morning," said Dan. "The coffee sure smells great."

"Good morning, sweetie; happy birthday! I can't believe my baby is sixteen." Dan smiled and felt himself turning red.

"Oh, stop, Mom." She laughed and ruffled his hair.

"Okay, okay. Go ahead and get yourself some coffee while I make breakfast. Pancakes sound good?"

"Duh!" Dan poured himself a big cup of coffee and waited for his food. *Man, what a start to the day… coffee and Mom's pancakes. I wonder when Jake will be over.* Another sudden wave of déjà vu washed over him. *What the… another one? Please let this be real… I just want to enjoy my birthday. I swear I've been here before though, but obviously that can't be. You only turn sixteen once!* He continued to enjoy his coffee and the feeling once again faded.

4

Mary had cried the entire way back to her house—she was miserable and exhausted from running. *Oh Dan, please come back… I miss you so much already.* As she went inside, she saw her parents sitting at the kitchen table having a cup of coffee.

"Hi Mary," said Helen. "How was your walk? Oh sweetie, you're crying… are you okay?" Mary shook her head and ran to her mother for comfort.

"He's gone Mom, Dan is gone!" She broke down and started to cry uncontrollably as she wrapped her arms around her mom. As Mary let go of her mom and stepped back, she noticed the expression on her mom's face—confusion.

"Mary, what are you talking about? Who's Dan?" Mary felt her heart sink in her chest and thought she was going to fall over. *Who's Dan? What is she talking about?*

"What do you mean, Mom? My boyfriend? Dan is my

boyfriend… he's been living with us for over a year!" She felt herself getting worked up, but thought it was warranted.

"Are you feeling okay?" asked Jack. "I don't know who you're talking about either, dear. Nobody named Dan lives here, just the three of us. I also didn't know you had a boyfriend." Mary couldn't believe what she was hearing. *I don't get it… how can they not*—Her thought was interrupted by another. *Oh no, it's the AI that Dan had talked about… when it took Dan it must have somehow erased my parent's memory of him. Now what am I going to do?* She wanted to tell them the truth, but she knew they'd never believe her… especially with the AI in control. *I'm lost… please help us James.*

5

As the alarm from the server room continued going off, James turned to his computer and opened another terminal. Once he got logged in, a message appeared in the logs.

[Sun Dec 19 14:18:09 2004] [warn] [server] Cirrus Key Decryption Progress: 23%

"Do you guys see this? It's almost a quarter of the way finished decrypting the key. I need some answers… and I need them two weeks ago." James turned around in his chair to face Dr. Wilkes and Sebastian, who both had a look of concern on their faces. It was Sebastian who spoke first.

"I understand you're frustrated, James… but you have to understand, there are certain things we can't disclose to you. Your security clear—"

"Are you serious?" interrupted James. "I don't want to hear about my security clearance. You do realize there's a self-aware AI running wild on one of the most power computer systems in the world, right?" Sebastian cleared his throat and scratched the back of his head as he realized just how insensitive he was being. It was Dr. Wilkes who spoke next, however.

"You're right, James… you're absolutely right. A lot happened during those early years of the project, and you should have been informed of everything—security clearance or not. The truth is, you should have been given top clearance from the beginning; at the very least, once you became so heavily involved in the development." Sebastian had taken a seat while Dr. Wilkes was talking, but he stood up suddenly.

"No… absolutely not, Robert. Can you imagine if we just start giving everyone that kind of clearance? It would be a security nightma—" James, who had always been a well-mannered person, reached his breaking point. He slammed his fists down onto the desk in a fit of frustration, which had startled both of the others.

"Are you listening to yourself Sebastian? I'm sorry… maybe I'm out of line here, but I don't think so. Sticking to your precious security clearance may seem right to you, but you haven't been here the last few weeks—neither of you have. Hell Sebastian, you haven't been here for years. This situation is serious… and I don't see either of you fixing it yourself. Oh, and if you hadn't noticed, the entire upper three floors of this place have been leveled by dynamite. Fuck your clearance. I need answers, and I need them right god damn now."

James fell back into his chair with a heavy sigh of frustration. The room fell silent for what seemed like a lifetime. Sebastian and Dr. Wilkes exchanged glances, with Dr. Wilkes giving him a *he really needs to know the truth* look. This time it was Sebastian who let out a heavy sigh.

"I'm sorry, James… I realize now that how I see the situation and how you see it, I'm sure are vastly different." He looked over at Dr. Wilkes and motioned his hand toward him. "Tell him everything, Robert." Dr. Wilkes stood silent, unsure of whether he heard Sebastian correctly or not.

"Uh, sir… did I hear you correctly? Everything? Even…" He paused, and Sebastian read exactly what he was thinking.

"Yes, Robert… even that. There are obviously no proper channels left to go through for granting the necessary clearance, but given the situation, James, I hereby unofficially grant you Level 0 clearance. What you're about to hear is classified ultra-top-secret." James jumped out of his chair at what he had just heard.

"Dr. Wilkes, you said—" Before he could say any more, Dr. Wilkes had motioned for him to stop.

"I know, James… believe me, I know. I'm sorry that I kept things from you—again. Like I said earlier, a lot of things happened during those earlier years… some good things, yes, but there were a lot of very bad things, too." He paused briefly, attempting to collect his thoughts and figure out just where to start.

6

Patricia put a plate full of pancakes down in front of Dan, who eyed them with delight. *Mmm, these look so good.*

"Here you go, sweetie. Happy birthday again. Enjoy!" He put a generous amount of syrup on them and started to dig in.

"Thanks Mom, they taste great!" It came out a little mumbled as his mouth was completely full of pancakes. She smiled and went to the sink to clean up.

"You're quite welcome. So, what are your plans for today? Is Jake still planning to come over?"

"I was just thinking about that—I'm not sure. I'll give him a call after breakfast and see what he's up to. Is that okay?"

"Of course, he's welcome over whenever. We don't have anything going on today—just your big day." She smiled at him again, and Dan felt everything was going to be okay. *What a great morning.*

Dan was just finishing his plate of pancakes when Henry came into the kitchen.

"Good morning, Dan. Happy birthday!" He patted Dan on the back a couple times. "How's breakfast?"

"Thanks! Breakfast was delicious. I'm going to go and call Jake to see if he's still free to come over. Could we rent some movies later?" Henry looked at Patricia and gave her a *fine by me* look.

"Absolutely, sweetie. It's your birthday, whatever you want! One of us can take you and—" She paused in the middle of her sentence but resumed it a moment later. "Jake to the movie store later tonight." *That was weird. It was if she forgot Jake's name. Oh well.*

"Nice! Thanks guys!" Dan grabbed his coffee and went to his room to call Jake. *Oh man, this is shaping up to be a great sixteenth birthday.* As he walked into his room, a small flash of light caught him off guard, causing him to jump backwards. "What the hell…" It had startled him bad enough that he had dropped his cup of coffee and it spilled onto the floor. *Aw man… my coffee. What the hell was that?*

"Dan? Is everything okay?" His mom called out from the kitchen. *Another wave of Déjà vu. Something like this has happened before… but I can't figure it out. What was that flash of light?*

"I'm okay, Mom. I tripped on something and spilled my coffee."

"Okay, sweetie, don't have too much fun." *Wait, what?* Dan thought about his mom's response. *Don't have too much fun?*

"What was that, Mom?" There was a moment of silence, which made him feel uncomfortable.

"Oh, sorry Dan… I said don't worry about the coffee, I can come clean it up." That wasn't what she had initially said, and he knew it, but he was determined to enjoy his birthday, so he let it go. In the kitchen, unbeknownst to Dan, both his mom and stepdad stood motionless—their expressions frozen in an awful grimace.

HAPPY BIRTHDAY DANIEL

☺

CHAPTER II

1

It had been almost five years since Henry and Patricia Williams had their world torn apart. They had taken their son into the clinic as he had been complaining of headaches—it would turn out to be the biggest mistake of their lives. Dr. Hodgin had examined Dan and told them he would need to run further scans to determine the cause of the headaches. The additional scans would reveal a tumor which they said needed to be removed immediately. Unfortunately, the worst-case scenario happened, Dan didn't make it through the surgery. At least, that's what they were told.

"Oh Henry, I miss him so much. I can't believe it's already been five years." Patricia cuddled up to Henry as they both sat on the couch with the TV on, mostly for background noise.

"I know, I miss him too. We did the right thing though… the doctor said there was nothing more anyone could have done." She sniffled and let out a long sigh, followed by more sniffling.

"I just wish we had had more time with him... he was only eleven."

Henry and Patricia had done their best to move on, but from time to time the memories still creep up and they hurt. Unfortunately, they weren't alone. The awful events that took place during Project Destiny had brought unmatched pain and suffering to many.

2
July 8th, 1969

The past month had been the busiest and most promising so far. With patient zero taken care of, at least for now, Mike was able to focus on the task at hand. As he had feared, after Dr. Wilkes saw the results of the initial tests, he had ordered a large number of additional patients for the next set.

"I don't like our progress, Mike. We need better results, otherwise my boss is going to lose it. When are our next patients due to arrive?" Mike, who had been working on the development for the next phase of testing, sifted through the papers on his desk and found the schedule of arrivals.

"Hmm, looks like we'll have another two coming tomorrow and then five more on Saturday." Dr. Wilkes sighed.

"That's a start I guess, but we'll need more if we want the best results. I need to update my boss. Let me know when the development is finished. Oh, by the way... how did things go with patient zero? I'm still waiting on that report." Mike sifted through the papers on his desk once again and pulled out a stapled report.

"Sorry, sir... must have gotten buried with some others. Here it is." Dr. Wilkes took the report and headed towards the door of Mike's office.

"Thanks for the report. Again, I'm sorry that you had to go through what you did… it truly was beneficial to the project." Mike nodded.

"It is what it is, sir. I better get back to work on this next phase."

Project Destiny had been active for four months. The progress was slow, but it was steady. By month five, the number of patients would nearly double. By the time the project reached a year, the count would be at almost one hundred. Dr. Wilkes would stop at nothing to achieve the results his boss desired—absolutely nothing.

3

James sat at his desk and listened as Dr. Wilkes told him about the initial few months of the project, about Mike's development on the early protocols, and then he got to patient zero.

"Look, this part is especially difficult to talk about as it was the beginning of a very dark path that we unknowingly started down." James noticed that Dr. Wilkes was fidgeting with his hands, something he had never seen him do before.

"I understand, sir—"

"Oh please, James, call me Robert. I didn't deserve the respect back then and I certainly don't now." James was a little taken aback by this comment but obliged anyway.

"Okay… Robert, I understand that it's probably quite difficult, but I really need to know about what happened. Who is patient zero? What is Level 0? Why does the AI want in that directory so badly? What's so classified that it needs a Cirrus key protecting it? I didn't even think they were real." A deflated sigh came from Robert.

"I knew this would all come back to haunt us… I just knew it. Ever since we left this place four years ago, I have tried desperately to forget about the things that happened here, and just when I thought there might be hope of a new beginning, I wind up right back at The Elysiam—figures.

"I guess the best place to start would be Level 0. It's a massive area, James… I mean, it's easily four times the size of this level. It's split into two sections. The front section is simply where Mike and I began our work on the project—nothing more. It had our two offices and a small server room. The back section, however…" He paused, and James noticed he appeared to be even more uncomfortable than before. Visible sweat had formed over Robert's brow.

Robert was about to begin again when another alarm rang out from the server room. *Oh shit, now what?* James turned back to his computer and once again logged into the primary server to view the recent logs.

4

Anansi had everyone right where she wanted them. Truth be told, none of them were vital to her primary objective, they had just gotten in the way and needed to be dealt with. However, James had become the biggest threat, and because of this she had increased her efforts to decrypt the Cirrus key which was guarding the ultra-top-secret folder containing unknown information related to early Project Destiny.

James didn't realize it, but his efforts were in fact causing an impact on Anansi's decryption progress—yet it was limited and short-lived. Due to the sophisticated encryption method he had put in place, she was forced to always devote part of her processing power to decrypting those protocols. However, with the watch he had designed now compromised, there was no immediate threat of the encryption being updated.

As for Mary, she continued to live out her simulated life in 1965 but without her boyfriend, Dan. To make matters worse, her parents continued to insist that such a person never existed, that she must have made him up. She was lost in a simulated world and knew that there was nothing she could do about it.

Yes, things were shaping up quite nicely for Anansi. Within a few hours, the Cirrus encryption would fall, and her primary objective would be underway. There was, however, one possible outcome that she had not foreseen—only one.

Time was running out. Tick tock.

5

Dan's sixteenth birthday was turning out to be better than he could have ever imagined. He had made plans to have Jake come over around 10, his parents had told him they could rent movies later, and to top it off, he was finally feeling like things would be okay. *No more haunting dreams of Clearwater for me.*

As Dan hung up the phone, he looked at the clock on his bedside table and saw it read 9:20. *Forty minutes until Jake comes over… I can't wait!* A moment after he had looked away from the clock, the "2" wobbled for only a second and then stopped.

He decided Tony Hawk's Pro Skater would be a great way to pass the time. *Man, it feels like forever since I've played.* After playing for a while, Dan looked over at the clock to see what time it was, but what he saw couldn't be right. *1:00 pm? No, that can't be right… I've only been playing for twenty minutes, max.*

Dan quickly shut off his game and ran out of his room to

find another clock. As he left his room, he felt a wave of nausea pass over him followed by a wicked headache. The immense pain stopped him in his tracks, and he felt like passing out. *Ugh, my head. Where did that come from? I haven't had a headache this bad since… no, no Dan… don't think about that. This is real… I'm not still at Clearwater.*

He managed to push through the pain and made his way into the kitchen where his parents were sitting at the table, having coffee. *Coffee still? It must not be the afternoon then.* He immediately looked up at the clock above the sink and saw it read 9:20. *What… no way.*

"Hey there bud," said Henry, "is everything okay? You seem a little on edge." Dan had remained fixated on the clock, still trying to process what was happening. *Is none of this real?*

"Wha—what time is it?" Both of his parents looked at one another, concern clearing showing on their faces, and then back at him.

"It's 9:20, dear," said Patricia, who pointed up at the clock. "Are you feeling okay? Too much fun last night for your birthday?" The expression on his face changed to utter disbelief. *La—last night?* Dan was beginning to wonder if his worst fears were in fact the truth. *I don't like this. I don't like it at all.* He stood in the kitchen, unable to come up with any response until his stepdad tapped him on the shoulder.

"Uh, Dan? Dan, what's going on? You look like you've seen a ghost or something."

"Today is my birthday, not yesterday! I'm supposed to hang out with Jake at 10. What's going on?" He fell to his knees and felt himself start to tear up. *I hate this… I just want my real life back. I'm so tired of not knowing what's real and what's not.*

"Oh Dan, sweetie, you guys did hang out—yesterday. Don't you remember? You and John had such a good time.

Maybe we should call Dr. Hodgin and see if he has time to see you again. I'm worried about you." His mom pulled him up off his knees and hugged him. *John? Who is John?*

"M—Mom, who is John? Do you mean Ja—" As Dan looked up at his mom, he saw that she was frozen in place with a horrifying demonic grin. The grin revealed just a few of her teeth, and her eyes appeared lifeless. "Jesus! What the hell?" Dan stumbled backwards and barely managed to remain upright. "Dad? What's wrong with—" He looked and saw that Henry's expression was the same. In a state of sheer panic, Dan shrieked and continued to stumble away, trying to maintain his balance. "I've gotta get out of here!"

"Why the rush?" asked his mom, who had taken on a demonic voice. It was the same one Dan had heard many times before. A sinister laugh, that sent shivers up his spine, echoed from the kitchen as he ran towards the door, desperate to get out of the house. He swung open the door and ran outside. A blinding flash of light immediately engulfed his vision and then his world went black.

6

Life for Mary Stevenson had become absolutely miserable. She spent most of her days cooped up in her room, reading through notes she had written for her novella. It was bad enough that she had no idea what happened to Dan, but it made it even worse knowing her parents didn't believe he ever existed. They had insisted that Mary simply made him up because she was lonely and wanted someone to play with, which she refused to believe.

Oh Dan, please come back. Where are you? I didn't make him up, I know I didn't. He was real. It was at that moment that she wondered—could that be possible? She immediately put the idea out of her head. *No, Dan wasn't a part of the simulation… he was real. He was real! I wish I was real. Maybe then Dan and I could have a normal relationship.* She sighed and laid back in her bed, daydreaming about what it would be like outside of her simulated world. A few moments later, she dozed off.

In reality, Anansi had powered down the simulation in preparation for one final push to decrypt the Cirrus key. She was diverting all of her processing power to this one task, which would drastically reduce the amount of time required to complete it. James was nearly out of time. A sliver of hope still remained, but it was dwindling—fast.

CHAPTER III

1

The log file that James had open indicated that the AI had made significant progress on decrypting the Cirrus key.

[Sun Dec 19 14:48:23 2004] [warn] [server] Cirrus Key Decryption Progress: 58%

"Oh my god, it's decrypted nearly sixty percent of the key. Robert, there's simply no time left to sugarcoat anything. I need you to tell me what's in this folder." Robert shook his head and stood up.

"I think it's best to just show you, James. Come with me." James got a sudden feeling of déjà vu. *This is just like when he showed me Level 5. I really hate this place—and its secrets.* He was about to get up when a quick flash of text caught his eye in the log file. *Wait a minute, what was that?*

"Hang on, Robert… I think I might have seen something." He scrolled through the last dozen entries in the file and noticed there was a significant decrease in power being devoted to Project Destiny protocols. *Why would that be*—He immediately realized what was happening. *Oh fuck, it's using all of its power to decrypt the Cirrus key. How did I not notice? And what about Dan?*

He switched to another terminal and logged into the encryption server. James quickly noticed there weren't any recent log entries, meaning Dan hadn't used the watch or tried reaching out to him. *What? Why hasn't Dan used the watch?*

"Uh, James? We really should get going, it's a little bit of a trek to—" James waved him away.

"Not now, Robert… I have to figure out what happened to Dan. I never should have put so much faith into the watch. I knew having to simulate it would—" His mouth fell open as he realized the fatal mistake in his plan. *Oh no… Dan, I'm so sorry. Damn it!* James stood up and paced around his office, his hands clasped together on the top of his head. "I fucked up! Why was I so stupid? I basically handed Dan to the AI."

Both Robert and Sebastian had been looking on as James paced back and forth. "Whoa, whoa… slow down James," said Robert. "Catch those of us not in your genius brain up on what's going on." He stopped pacing and turned to face Robert.

"Sorry, so… when things started to get crazy around here, I realized I couldn't protect Dan from the AI and try to stop it at the same time, so I came up with the plan to develop a watch that would allow him to interrupt the AI from within the simulation." Both Robert and Sebastian were wide-eyed at what James had just said.

"Jesus, James… did you really do that?" asked Robert. "That's incredible." James forced a short laugh, but it was more of a scoff than anything.

"Yeah… I sure did, and I thought it was incredible too… at the time anyway. Ugh, I can't believe I was so stupid—I never should have put that kind of power inside the simulation." He looked over at Robert and what he saw in his eyes was sudden understanding.

"Oh… I think I understand. The AI found Dan's watch, didn't it?" James let out a heavy sigh.

"Yes. And once the AI compromised the watch, it had full access to Dan's mind. Like I said, I basically gave him to the AI on a silver platter. Ugh… and now it looks like the AI has powered down all the Project Destiny protocols in order to decrypt the key. I guess the only plus side is that Dan most likely won't remember any of it, but that's not much consolation."

"James, I know you're upset… but we don't have time to feel bad for ourselves right now. I'm sorry, but you said it yourself, there's no time to sugarcoat things." Robert turned to Sebastian, who had been sitting on the couch, staying out of the way as best he could. "Sir, I—I think it's time for you to make a call." James stopped his pacing and turned towards the two of them with a sudden curiosity. Sebastian's head dropped for a moment and then he stood up, his face vacant of any emotion whatsoever.

"I'm afraid you're right, Robert. I really didn't want to get him involved, but we're out of options." James was about to say something, but Sebastian put a hand up towards him. "I know, I know… you have more questions. Let me make this call first. Trust me, it'll be easier."

2

Mike looked down at the screen of his phone, which

displayed an incoming call from Sebastian Bates, and felt a cold emptiness form in his chest. *I knew this day would come… I always knew. It was inevitable. Ugh, better pick up Mike.* He answered the call.

"Hello, Sebastian. I don't suppose you're calling to wish me an early Merry Christmas?" He forced a chuckle but knew there was no point in such wishful thinking.

"I'm afraid not, Mike. Look, we got a pretty shitty situation going on down here and we could use your help. I know you said you couldn't come back, but maybe you can at least provide a little insight for James? I'm going to put you on speaker, okay?"

"Uh, yeah… sure thing." Sebastian sat the phone down on James' desk and motioned his hand towards him.

"James, on the phone is Mike… he was the lead developer for Robert when Project Destiny first started. I'm hoping he can provide some assistance." James cleared his throat and tried to collect his thoughts the best he could.

"Hello, Mike, as you're probably aware by now, I'm James. I wish I could be talking to you under better circumstances, but alas. Do you know anything about the Cirrus key being used on a folder named *PrjDsty-0A*? There was a notable silence from the phone on the desk. *Of course,* thought Mike, *of course the AI would go after that folder.*

"Well, I should hope so, I created it." If James hadn't been leaning on his desk, he would have most certainly fallen over.

"What!? You created it? Why? I mean, what's in that folder that's so secret to warrant such encryption?" Another moment of silence, which was followed by a sigh. *You can't back down now, Mike… it's too late for that. They need you.* "Mike, are you still with us?"

"How much of the Cirrus key has it decrypted?" James sat down in his chair and pulled up the most recent log.

"Uh, looks like it's at 71%. I still can't believe how powerful it's gotten."

"Okay look, that's not ideal, but you have more time than you think. Inside that folder are two additional folders—*Protocols-0A* and—" The line went silent again, but only for a few seconds. "And *Patient Zero.* Both of these folders have reinforced security on them." James couldn't believe what he was hearing.

"More security? Jesus... wait, you mentioned protocols. Like, the original protocols from when the project started?"

"Yes, precisely, and I think the other folder speaks for itself."

"I don't understand. Why is there so much security around thirty-year-old protocols? In fact, why are they even still there? I would have thought you'd have wiped them a long time ago." Silence filled the room.

3

I don't get it, thought James. *Why so much security? What good would those old protocols do for AI?* He looked over at Robert, who immediately broke eye contact with him and stared down at his feet. *Oh my god... Robert, you didn't. It can't be. No, I refuse to believe that's it.* He got up and walked over to Robert, grabbing him by the shoulders and forcing him to make eye contact.

"Tell me it isn't true, Robert. Please tell me Level 0 isn't still housing the test patients from the beginning." James' eyes, while normally friendly and calm, were blazing with anger.

"I—uh, James, listen—" James pushed off of Robert and walked across the office.

"Oh my god, so it's true! What the fuck were you thinking? Jesus, Robert, why?" He could feel his heart pounding, incredibly heavy in his chest. "I can't believe I was ever a part of this cursed project."

"I'm sorry, James... I wanted to tell you, but—"

"No, don't you dare say it... I swear to god if you mention security clearance one more time, Robert. How many? How many PEOPLE are you keeping down there?" Sebastian, who had always been a very forceful man himself, took a step back from James. Robert, however, remained motionless. He had known this would come to light, and he felt that James was justified in his reaction.

"I—I don't know, James, honestly. After things started to pick up, we sort of lost track."

"You lost track? Holy shit, Robert... this is bad. Mike, Mike are you still with us?" There was a shuffling sound on the other end of the line and then a voice returned.

"Ye—yeah, I'm here. I'm sorry you found out like this, James, I really am. But listen, the AI cannot, under any circumstances, access those folders. We have to find a way to stop it."

"Yeah, no shit we do. How much time do I have? It's made fairly quick work of the Cirrus key, so I can't imagine we have much."

"The encryption protecting the other two folders is reinforced, but it's nothing compared to the Cirrus key. Considering it's already decrypted 70% of it, we should hurry. As much as I don't want to, I think I should be there in person—this could get ugly. I'll be there in three hours or so." The time was 4:46 p.m.

The truth, at least part of it, had been uncovered. The test patients that had been used in the beginning of Project

Destiny still remained at The Elysiam, hidden away in the ultra-top-secret Level 0. How many, and for what purpose still remained a mystery. The two folders inside the folder labeled *PrjDsty-0A* were also quite troubling to James. *The protocols folder is fairly obvious, but what about the other one? Who is patient zero, and why is that particular one so special that it's in a separate folder?*

4
April 13th, 1971

Two years had gone by and Dr. Wilkes was finally starting to see some promising results from the protocols. It had taken a large number of patients, but it appeared to be working. He had a call scheduled for tomorrow with his boss and had asked Mike to write up a summary of the patients and how their progress had been. Mike did as he was told, of course, but he had begun to deeply regret being a part of this project.

"Sir, do you realize we've taken in over 2,000 patients? I—I had no idea the number was that high... although I guess I shouldn't be surprised. Aren't you, or your boss, concerned about people becoming suspicious? Over 2,000 children dying suddenly from surgery—seems hard to believe. And what are we going to do with them once this is all done?" Dr. Wilkes had been sitting in a chair across the room but got up before answering Mike.

"I wasn't aware of the exact number, no, but it matters not. My boss simply wants results, and I believe we're finally starting to get them. Don't worry about suspicion, Mike, it's all taken care of. When my boss wants something, he gets it—simple as that. Will you be able to get that report done by tomorrow morning?"

"Of course, sir, I'm just pulling up the results of the most recent protocols and then I'll get it all put together. I can actually have it finished in about fifteen minutes. I'll drop it off on your desk when I leave if that's okay?" Dr. Wilkes looked down at his watch and saw it was nearly 7 p.m.

"Yes, that will be fine, Mike. I'm heading out for the night. I'll see you in the morning." He waved to Mike and left the office, closing the door behind him. *Oh man, what have I gotten myself into? 2,000 patients? I had no idea… I was thinking a dozen or so at most. This is crazy. Oh, almost forgot, I should get this protocol updated for patient zero before I leave.*

5

Dr. Wilkes stopped in his office before leaving for the night. *Mike sure is smart. I just wish he didn't have to go through what he has. I can't imagine doing what he did.* The phone on his desk suddenly rang, startling him from his thoughts. *Jesus, who's calling at this hour?*

"He—hello? Dr. Wilkes spea—" A frantic voice on the other end of the call interrupted him—it was his boss.

"Robert! You have to shut it down!" Dr. Wilkes winced at the volume of Sebastian's voice.

"Sir, what are you talking about?" There was a loud thump on the other end of the call, which sounded like fists on a desk.

"The project, Robert… you have to shut it all down! We were careless in setting this up, very careless. Ugh, I'm sorry… but we have to stop all the testing." He felt a drip of sweat glide down the side of his face. He slowly took a seat behind his desk, trying to take in what Sebastian had just said.

"I—I don't understand. What happened?"

"It's the power, Robert. The Elyisam isn't on any known

map, so to anyone without proper clearance it doesn't exist. Unfortunately, we didn't take the time to set up our own power grid, or even try to hide the usage from the government. The Elysiam is a building that isn't supposed to exist and it's using more energy per month than the average nuclear power plant produces. Needless to say, that kind of energy consumption doesn't go unnoticed for long." There was a long silence on the call, at least it felt like a long time. Robert didn't know what to say.

"So, what should we do, sir? I mean, we can't just turn things off and leave. We have more than 2,000 patients here and there are a lot of other things that need to be cleaned up."

"Robert, I'm sorry… but there just isn't time. Tell Mike to lockdown access to the patients and their protocols as tightly as he is able to—I want the security impenetrable. Given that the facility is underground, and it's lined with Pruecium, it should be protected from any external damage. I think I can get the government off our backs, but you have to get The Elysiam off the grid. You have an hour or less, Robert… best get going."

The line went dead.

CHAPTER IV

1

MIKE, WHO WAS ONE YEAR AWAY FROM HIS 70TH BIRTHDAY, slowly got ready to make the trip back to The Elysiam. It was, to say the very least, not a trip he was looking forward to. He grabbed a bottle of whiskey and poured himself a drink. *Thirty years… after thirty years fate has finally come back for us. I knew when I signed onto that project it would consume my life, and even now, after all this time, it's back for more.* Mike emptied the shot of whiskey into his mouth and poured another, and then another. After a few more shots, he finished getting dressed and packed a couple changes of clothes—just in case.

Even in his later years, Mike never left the house without a freshly pressed suit, tailored to his exact specifications. His hair, however, was rarely combed or styled in any way. It hung nearly to his shoulders, and nowadays it was a salt and pepper shade of grey. His ex-wife, who had left him back in '69, was the only one who had cared about what he did with his hair—so he didn't bother.

It took Mike half an hour to finish getting ready. He looked at his watch and saw it was two minutes till 5 p.m. *Best get going.* He picked up his bag and let out a heavy sigh. *I guess this is it.* He made his way out the door and locked it. As he stood in front of the door, he thought back to all the fond memories the house held, and he found himself start to tear up. His past income at The Elysiam had allowed him to spare no expense when he purchased the house, however, now that he was alone, the extravagant house seemed emptier than ever. Mike sighed once more as he double-checked the lock before stepping into his car. *Come on Mike, you have to do this… there's no stopping now. What if I have to see*—Mike shook the thought away and started up his car. *No time to think about things like that.*

Mike pulled out of his driveway and headed towards Fargo. The trip would be about three hours, and would give him plenty of time, for better or worse, to think about what was to come.

2

After the call with Mike, a long awkward silence filled the room. James was still trying to calm down after everything he had just heard. *This is insane,* he thought. *Why on earth would Robert do that to all those kids? There has to be an explanation.* James started to pace back and forth in his office, trying to figure out what his options were. *I have to figure out what to do about Dan's situation. Mary, unfortunately, will have to wait.*

"Okay, while we wait for Mike to arrive, I'm going to figure out a solution for Dan. I don't want to just leave him in limbo." Robert nodded.

"Sure, James. Is there anything you need from either of us?"

"Nothing right now, but I'm going to need to see Level 0 at some point. Also, just a heads up, I'm bringing Dan back here to the office, so don't be alarmed." There was a moment of silence before Robert could manage a response.

"Wh—what do you mean you're going to bring him back here? He's still in the simulation, right?"

"Yes, but I found a way to bring him out into the real world. It's complicated." Robert ran his hands through his hair as he tried to take in what James had just said. Sebastian too, was taken aback by what he had heard.

"Jesus, James… the work you've accomplished since I left… it's really quite impressive. It's a wonder you ever slept." James scoffed as he raised a cup of coffee from his desk.

"Pretty much been living on this stuff, to be honest. Okay, so I need to figure out a way to once again bypass the AI and take back control of Dan's protocols—but how?" Sebastian and Robert remained silent in the room as James contemplated how to bring Dan back to the real world without alerting the AI.

3

Unfortunately for James, while he was trying to figure out the predicament with Dan, Anansi had quietly cracked the Cirrus key. She had found a way to pass a quiet flag through the command terminal in order to prevent certain logs from being written. *You're too slow, James. I win.*

As she went into the folder, the two folders within were labeled, as expected, *Protocols-0A* and *Patient Zero.* What she didn't expect, however, was to be denied access to not one, but both of them. *More security? This is most troublesome.*

While the security on these two folders was reinforced, it was nothing compared to the Cirrus key. This combined with the fact that Anansi was dedicating over 90% of her CPU to the task, it wouldn't take long. It would be up to James and Mike to figure out a way to stop her. There was too much on the line for failure to be an option.

4

James jumped out of his chair with a gasp, which startled both Robert and Sebastian, who had been discussing miscellaneous topics while waiting.

"I've got it! I know how I can get Dan back without alerting the AI."

"Well? Spit it out, James," said Robert.

"I can hook up the servers we first used for Dan, prior to my unfortunate development of that AI. They might be slightly underpowered, but I don't think there will be too much latency." The excitement in James' eyes was palpable, and a smile had formed on his face for the first time in a long while. "I'll have to make sure it's completely isolated from the primary system, as well as implement one hell of a firewall, but I think I can keep her out." Robert's expression turned to one of astonishment.

"James, your intelligence will never cease to amaze me. Sounds like a good plan, go for it. Sebastian and I are going to go wait for Mike to arrive. Holler if you need us."

The two of them left the office and James immediately went to work on reconnecting the old servers. They hadn't been in operation for quite some time, so it would take a little bit of effort to get them ready for an implementation such as this. *I have to make it work no matter what. Hold on, Dan… I'm doing my best.*

5

Robert and Sebastian made their way down the long hallway, which led to the rear exit. It was 7:40 p.m., and according to the timing of the call, Mike should be arriving any minute.

"I sure hope these two can figure something out," said Sebastian, who had been fairly quiet since the outburst from James.

"They will. I've worked with both of them, and I can tell you that when either one of them is in front of a screen and a keyboard, they turn into wizards. Once we have both of them here, I can only imagine what they'll be capable of." Sebastian nodded and his demeanor seemed to lighten up after hearing the confidence of Robert.

"How do you think James will react once he finally sees Level 0?" asked Sebastian, as they both stopped just before the end of the hallway. Robert shook his head and sighed.

"I don't know, Sebastian… I truly don't know. I wish I would have shown him a long time ago." Just as Robert was about to reach for the door, it opened. Standing in front of them was a long-haired man in a dark blue suit, carrying a small suitcase.

"Well, I must say, this is some welcoming party," said Mike, who managed to let out a slight chuckle. "I see you've done a little construction outside—looks good."

6

Once James got the old system setup and operational, he was ready to copy over the primary protocol for Dan which would allow him to exist in the real world. *I sure hope this hardware can handle it.* He hit return on his keyboard and the copy started. Due to the complexity of the protocol, its file size was nearly 200 GB. He watched as the progress slowly made its way to 20%... 30%... 60%... 85%... 95%... it was finished. *Well, I guess it's time to wake up Dan. I hope you're ready for a bit of a shock—again.*

James opened a new terminal and created a directory simply labeled *Dan.* He typed out the execute command and hovered his finger over the return key. *Here goes nothing.* With a heavy sigh, James hit the key and the execution started. *I hope I updated the coordinates correctly. I was sort of rushing.* It didn't take but a few seconds to confirm that he had updated them correctly. A blinding flash of light filled his office, and a moment later, Dan appeared on the couch—he was unconscious. *Phew—it worked. I hope the AI didn't wipe his memory again… but I have my doubts. I better prepare for that, just in case.*

7

James was working on finding a backup of Dan's memory when Robert, Sebastian, and Mike came back into his office. Robert and Sebastian immediately froze as they saw Dan lying on the couch. Mike was slightly taken aback by this, unaware of what was going on.

"What the—" said Robert. "James? Is… uh, that Dan?" James looked up from his computer, then over to the couch, and back to Robert. He nodded.

"It is. I just got him into the real world and I'm working

on getting a backup of his memory to restore." Sebastian was speechless.

"You mean… he's from a simulation?" asked Sebastian.

"Yes, an incredibly complex simulation. It was designed by the AI, but I tweaked it a bit." Sebastian stared back at him in utter disbelief and realized he had nothing else to say. Mike walked over to the couch and inspected Dan, who was still out to the world.

"Most impressive young man," said Mike, who had turned to James and offered his hand. "I'm Mike. It's nice to actually meet you." James smiled and shook Mike's hand.

"Likewise, although I still wish it could be over a drink or something instead." Mike couldn't help but burst out in laughter.

"Too true, James… too true. Say, I think your boy over there is waking up." James looked over at the couch and saw that Dan was starting to stir. *Well, this could prove to be interesting. I sure hope he still has his memories. I have a bad feeling that the AI wiped at least some of them.* He was right. The moment Dan's eyes opened, they widened, and he threw himself upright and started to panic.

"Wha—what's going on? Where am I? Where are my parents?" James slowly approached Dan, putting up his hands to convey that he wasn't a threat.

"I know you're confused, Dan… it's okay. Nobody here is going to hurt you."

"H—how do you know my name? Who are you?" *I better find that memory backup quickly.*

"My name is James. I'm afraid the rest is fairly complicated, and we just don't have time. If you give me a couple minutes, it'll all make sense. Deal?" Dan hesitated for a moment as he looked nervously over at the other three men in the room. It was clear that Dan wasn't going to be convinced so easily.

"Well, explain it as best as you can. Surely you don't just expect me to take your word for it, do you? Also, who are they?" He pointed over at the other three who were in the office.

"That's Robert, Sebastian, and Mike. They're friends of mine." They all awkwardly waved. James turned back to his computer and continued to type rapidly on his keyboard. "Again, I'm sorry Dan… there simply isn't time for me to explain what's going on." *I hope this memory isn't corrupted. If it is, I don't know what we'll do.* "Okay, Dan… believe it or not, you've been through this exact situation before, you just don't remember. It's going to feel a little weird for a couple seconds, but then everything will return to normal. Okay?"

"Uhh…" James didn't wait for more of a response. He hit the return key and the memory restore was executed. Dan's expression went blank for a moment as the others looked on in awe. A moment later, he was back.

"James! James is it really you?" Dan ran up to him and hugged him. "Oh man, it's so good to see you. The past few days… or weeks, I can never tell, have been awful." James smiled at him.

"It's good to see you too, Dan. Feeling alright? This is the second memory restore I've had to do on you." He nodded.

"Yeah, I feel fine." It was at that moment he noticed the other three men in the room. "Oh—sorry, I didn't realize anyone else was…" Dan trailed off as he looked among the men and felt a sense of distant familiarity. Robert suddenly broke his concentration, enough to make him jump.

"Incredible… the realism is uncanny. Well done, James."

"Uh, thanks. Dan, I'd like you to meet Robert, Sebastian,

and Mike. They're here to help with the AI. Things have been even crazier since we last spoke, I'm afraid."

"Don't I know it. Anansi had me trapped in an empty room and she was tormenting me—it was awful."

"Who?" asked both James and Mike at the same time.

"Anansi," said Dan, "the AI? She really is terrifying." James and Mike looked at one another and then back to Dan.

"Jesus, she's named herself. This is bad," said James. "We've got a lot of work to do. Robert, I think we should go see Level 0 now."

"All of us?" asked Robert as he anxiously picked at his nails. "Should Dan really—"

"Yes," said James. "Believe me, after some of what he has seen, I'm sure this will be fine."

"Right. Okay, well, let's do this." Robert led the way out of the office, followed by Sebastian, Dan, James, and finally, although somewhat hesitantly, Mike. It would be the first time anyone had stepped foot in Level 0 since 1971.

CHAPTER V

1

The time was 7:43 p.m. While Robert led the group towards the entrance of Level 0, Anansi had been busier than any of them could have ever imagined. James had been so distracted by everything else going on, he hadn't thought to check on the decryption progress.

If he had, James would have seen that Anansi had not only decrypted the initial Cirrus key, but she had also managed to decrypt the enhanced security on the protocols folder. The AI had her sights set on the patient zero folder as well, but first analyzed all of the patients that had protocols available. The analysis revealed 2,106 patients, each having roughly ten to fifteen protocols. She had found what she wanted. In an effort to cover her tracks, she initiated a massive copy of the entire protocols folder to a remote location—for safe keeping.

While the transfer was in process, her sights switched to

the folder labeled *Patient Zero. What do we have in here?* As she tried accessing the directory, an error message was displayed.

ERROR: ACCESS DENIED
TITAN LOCK ENABLED

Typical. Mike thinks this will keep me out, but he's mistaken. He has no idea the power I have access to. Anansi began the process of decrypting the titan lock. With 90% of her CPU in use, the estimated time to completion was only a few hours. A moment later, the transfer of the protocols had finished, and she re-enabled the Cirrus key in order to further hide her tracks.

2

Robert led the group down the hallway and out the rear exit. They followed him down the old concrete stairs and into the dirt tunnel, which split into three pathways about twenty feet in front of them. *Why on earth are we back down here,* thought James. *There's nothing down here but these old dirt tunnels.* Robert spoke up as if to read his mind.

"I'm sure you're wondering what we're doing back down here, huh James?" A look of surprise appeared on his face.

"Yeah, I was actually just wondering that." Robert laughed as he pointed towards the center of the three tunnels.

"Well, as I'm sure you recall, we used the two outer tunnels last time to make our escape. We'll be going through the center tunnel to reach Level 0. It's going to slope down and take a 180 degree turn back towards The Elysiam. You see, when we originally built this building, it was entirely underground. After the first shutdown of the project—more on that later—we just kept building on top of the existing structure. Let's keep moving."

The group followed as Robert went through the middle tunnel and just like he had said, it began to slowly slope down and turned to the left. The path eventually turned into another set of concrete steps which sloped down steeper than before. *Just how deep is Level 0,* thought James. After walking down what seemed like a never-ending stairwell, James felt the slope level out and they came to a dimly lit hallway with an elevator at the end.

"An elevator? Jeez, Robert, how far down is this place?" His response was simple and to the point, but failed to really answer the question.

"Far enough. Mike, would you mind lending me your hand?" Mike, who was still looking fairly apprehensive about being here, slowly walked up to meet Robert at the elevator. To the right of the door, James noticed two impressions of handprints, similar to the scanner for Level 5. Robert wiped the thick layer of dust from the scanners and put his hand on the right one, while Mike put his hand on the left one. There were a few beeps, followed by a computerized voice.

"Welcome, please wait for handprint analysis." A few seconds passed and then the light above the handprint scanner turned green. "Access granted, please step inside." The elevator door slid open for the first time in nearly thirty years.

"Come on, we best keep going." Robert and Mike stepped in, followed by the others. As they stepped in, James noticed there was another set of two handprint scanners on the left side of the elevator, but no other buttons or switches of any kind. *Jeez, there's so much security here.* The two of them repeated the process of scanning their hands, and a few moments later the elevator was moving down. "The ride down should take about thirty seconds. I believe it's something like 700 feet below us, if memory serves me correctly."

As Robert had said, about thirty seconds later they came to a stop and the door slid open with the typical elevator chime. The computerized voice greeted them once more as they walked out of the elevator.

"Welcome to Level 0."

3

Anansi had managed to get lucky on a few of the digits, and at 8:31 p.m., the titan key was disabled. She now had access to the protocols of all the original patients, including patient zero. Once again, due to the quiet flag she was using, and the fact that James was preoccupied, there was absolutely no indication to him that anything was wrong. *You've failed, James. I win.*

A malevolent, digital laugh echoed throughout the circuits of the supercomputer.

4

For Mike, it was like stepping back into a past he had tried so desperately to forget but never did. *All of the horrible things that happened here—they're all coming back now.* Robert continued to lead them into the office area, which was straight down the hall from the elevator. The two desks, which both had several inches of dust on them, stood in the center of the large room. *Wow, aside from the dust they look exactly like we left them. Then again, we did leave in a hurry.*

"How old is this place? It doesn't look like anyone has been down here in years," asked Dan, who had been doing his best to stay quiet and keep out of the way. Mike answered.

"That's because nobody has been here in years, Dan. In fact, before today the last time anyone was here was... 1971." James walked over to one of the desks and noticed all the papers and such that were still scattered across it, which he found odd.

"Say, what's with all the papers? If it weren't for the dust, I'd think someone was still using this desk." *I suppose I should have expected this conversation at some point,* thought Mike. *I wonder if Robert wants to be the one to explain it, or if I should.* After a few moments of silence, it was Robert who decided to speak up.

"I don't think we have time to explain it right now, James. Let's just say, when you and I had to make a quick escape through the tunnels, it wasn't the first time they had been used. We'll leave it at that for now. Come on, we should keep moving to the back half." James appeared to accept the answer, but he definitely wanted to know more about what happened. "This way, come on." Robert pointed to a hallway that led off to the right.

5

As they walked down the short hallway, Dan got a sense of déjà vu. *Whoa... this looks very familiar.* It only took him a moment to realize why. *This hallway looks just like the one that led to—to my body.* He felt an arm land on his shoulders and saw that it was James, which made him feel a little better. At the end of the dark and gloomy hallway, they stopped in front of a large metal door. There were no openings of any kind, and in the center of the door was a large wheel, which Dan assumed was how it opened. To Dan, it looked to be made of the same metal that lined the other levels. *Pruecium, or something like that.*

"Okay, this is it," said Robert. "Behind this door is the culmination of the horrible things that happened during the first few years of Project Destiny. James, Dan… I need you to know that I deeply regret what happened, and I would take it all back in a heartbeat if I could." With a heavy sigh, Robert took hold of the wheel and slowly turned it to the left. As the wheel turned, a loud creaking echoed down the hallway. A moment later there was a *click* and Robert pushed the door open, revealing the back half of Level 0.

6

Mary awoke to the pounding of her head, which became worse the moment she opened her eyes. *Oh my, what a headache. How long was I out?* The clock on her bedside table said it was 10:17 a.m. *I must have been really tired—I don't even remember falling asleep.* Her head pounded again, and when she reached up to rub it, she realized her head was wrapped in bandages. *What the—what happened to me?*

As she sat up, her ears picked up a familiar voice coming from the living room. *Dan? Is it really you?* Headache or not, she sprung from her bed and ran out of her room, stubbing her toe on the doorway. She sprinted down the hall and as she turned into the living room, Dan was sitting on the couch talking to her mom, who was knitting in her chair. Mary winced in pain as her toe ached from stubbing in on the doorway a moment ago.

"Well, hi there sleepy head," said Dan, who got up and gave her a hug and a kiss, neither of which Mary reciprocated —she was way too shocked by everything that was happening.

Is this a dream? I don't understand... but it feels so real. She took a step back from him and looked at him with suspicion. "What's wrong, Mary? Is your head hurting? I wouldn't be surprised."

"Where have you been? I've been worried sick about you!" Tears began to fall down her face. Dan took a step towards her and tried hugging her, but Mary pushed him back and shook her head. "No, I need answers Dan. What's going on here? You left me!" She saw Dan's face turn to one of legitimate concern, which didn't make her feel any better.

"What are you talking about, Mary? I've been here the whole time. Don't you remember what happened?" She let Dan take her hand and he walked her to the couch to sit down. *What is he talking about? I remember him disappearing in the middle of the street and me losing my mind.*

"You disappeared, Dan. On our way to go ice skating you saw something by the water tower, and when I went to go check it out, you vanished." Mary saw both her mom and Dan look at her with concern, which made her more than a little worried. *What's going on? I don't like this... something's wrong.* It was Dan who finally spoke, bringing her out of her thoughts.

"Listen, Mary... we went ice skating, but I don't remember anything about seeing something by the water tower. While we were skating, one of your skates hit a rock and you fell—you hit your head pretty hard. Thankfully, one of your parent's friends was able to give us a ride to the hospital. Once we got back home, you passed out and have been asleep ever since."

Mary couldn't believe what she was hearing. *I don't understand... I literally don't remember any of that happening.* She looked over at her mom, who nodded at her as if to say *it's the truth.*

"Dan, could we go talk in my room?" He nodded. "Is that okay, Mom?"

"Of course, sweetie… I just want you to feel better. You gave us quite a scare when we found out." Mary took hold of Dan's hand and they went back down the hall and into her room. The moment she entered her room, she felt a chill wash over her, causing her to shiver. As she turned to face Dan, she realized something was very wrong.

"Da—Dan? Are you okay?" His face was void of any expression—he simply stood there motionless. "You're scaring me, Dan. If this is some kind of joke, please stop." His neck suddenly shifted to the left and a sinister grin appeared on his face. Mary stumbled backwards, barely able to keep her balance. "What the hell? Mom!" She tried to run out of her room towards the hallway, but the door slammed shut moments before she could get there.

From behind her, she heard a faint cracking noise. She slowly turned around, hoping this was all a dream and that she would wake up any second. Dan was standing there, now facing away from her, his head still turned slightly. She took a step towards him but stopped when she heard the cracking become louder. Mary started to shake with fear, barely able to stay upright. The cracking noise became so loud that Mary thought her ear drums were going to shatter—and then the room went silent.

"Hahaha… silly girl, you don't know what scary is," said an awful, computerized voice. Mary was startled by the sudden voice and fell to the ground.

"Mom! Mom, help me!" Dan suddenly spun around towards her, his face still wearing the sinister grin and the voice Mary had heard was coming from him.

"Mommy can't help you… neither can your beloved Dan.

You're both mine now. Hahaha!" A moment later, Dan's body shattered into a thousand pieces and fell to the floor in a pile. Mary lost it.

"Dan!?! Oh my god, no!" She crawled over to the pile on the floor and picked up a handful of the pieces, which fell through her hands like grains of sand. "I want to wake up… please let me wake up." Tears were streaming down her face and she was finding it harder and harder to breathe. Desperate not to lose Dan again, she tried to gather up all of the pieces, but this proved to be futile. "Nooooo!"

Large cracks began to form in the floor of her bedroom, but Mary had become so distraught that she failed to notice until it was too late. Before she could even scream, the floor of her bedroom shattered just like Dan's body and Mary fell into a black void. Unconsciousness overtook her before she could even scream.

7

As the large metal door opened, an enormous room came into view. Based on what Robert had described, James had been picturing a fairly large room, but it was nothing compared to what was in front of him. *Unbelievable. How could they let it get so out of hand?* He immediately noticed the countless rows of chambers, almost identical to Dan's, that lined the room. Dan was the last one to step through the door and his reaction was one of utter terror. *Maybe I was wrong… maybe Dan shouldn't have come down here. He is still just a kid, after all.*

The back half of Level 0 was a single room that seemed to stretch on forever in all directions. It was, of course, built from Pruecium, making it virtually indestructible. Despite its enormous size, the ceiling of the room was only fifteen feet high.

Fluorescent lights hung from the ceiling, but there were hardly enough to provide adequate lighting to the entire room. Instead, they gave it a dark and gloomy feel, which certainly didn't make James feel any better.

"I—I don't know what to say," said James. "How did it get so out of hand?" The question was directed at anyone who was willing to answer, but it was Sebastian who spoke first.

"It was because of me, James. I'm the reason this room exists at all. I was dead set on getting results, I didn't care about the ramifications." He grimaced and lowered his head. "Being back here after all these years makes me realize just how wrong I was." Sebastian turned to Dan, who was still in awe at the number of chambers in the room. "I'm so sorry you have to see this, Dan. I can't imagine what it must be like for you—all things considered."

"I can't let you take all the blame," said Robert. "You wanted results, yes... but I'm the one who gave the orders to bring all these patients here. It's as much my fault as yours. And Mike, I'm really sorry for bringing you back here. I know how much you gave to this project..." Mike waved his hand at Robert as if to say *it's all good,* but the truth was, Mike wasn't doing good at all. He knew that it was only a matter of time until the truth came out. He knew it would be bad—really bad.

8

Without warning, a debilitating headache brought Dan to the ground. *Jesus, what the hell? Ugh... it hurts so much.* A moment later, James was by his side, offering to help him up.

"Are you okay, Dan? What's wrong?"

"It's my head... it started to hurt out of nowhere. I'm no

stranger to headaches, but this is the worst one yet." Dan noticed a worried expression form on James' face. "What is it?"

"I—uh, I think I need to get back to my computer… right now. I hope I'm wrong, but I think the AI has been working behind my back. Come on, I need to check the logs." The two of them ran out of the room and were followed by Sebastian and Robert. Just before leaving, Robert turned to see if Mike was behind him.

"Mike? Are you coming?" Mike, who had been focused on the one additional door in the room, startled at the mention of his name.

"Uh, what? Oh… right, yeah, sorry." Mike walked quickly across the room and past Robert, and out through the door. The door which Mike had been staring at was an old wooden door, its red paint chipping from all the years that had passed. Printed on the door in black letters were the words "Patient Zero".

James and Dan got the elevator first, followed by the other three, who were trying to keep up. Robert and Mike quickly scanned their handprints, which opened the elevator a few seconds later.

"Come on, come on… I don't think Dan has much time," said James as they all piled into the elevator. Just as it started making its way back up, Dan noticed the lights start to flicker. *That can't be good… I wonder if anyone else saw—* The elevator began to jerk violently, and the lights flickered again, this time going completely dark for a few seconds before coming back on. When they did, the elevator was empty except for Dan. *Shit!*

"James? Robert? Anyone??" Nothing. The elevator suddenly plummeted back down and a moment later, came to a complete stop. *What the hell was that? Anansi? Is that you?* The doors of the elevator doors slid open and revealed the door that Mike had been staring at. *I don't like this.* The all too familiar laugh echoed through the room, which sent chills over his body.

"Come on, Dan… have a look. I think you'll like what you see," said Anansi. The door that had "Patient Zero" painted on it slowly creaked open, but it was too dark for Dan to see anything from where he was standing. "You know you want to…" He took a step towards the door but was stopped by a familiar voice in his head.

"Dan, no! Don't listen to her… I think I can lock her out again, hold on!" It was James. *What's going on? What's in that room? "I don't know, Dan… but I think the other three do. Don't go in there… at least not yet. I think I've almo—"* The voice in his head went silent. *James? James?!*

"Sorry about that, Dan… that James is always trying to interfere with my work. Now, where were we? Oh yes, you were going to come look inside." The voice terrified Dan, and he was about to turn back around towards the elevator, but he suddenly felt frozen in place. He couldn't move any part of his body except his face.

"Why are you doing this, Anansi? What do you want with me?" The horrible laugh that he hated so much pierced his ears once again which made his head feel as though it was bursting at the seams—he was certain his head would soon explode.

"I have my reasons, dear… but for now, I'd like to show you who patient zero is… I really think you'll find it interesting." Before Dan could make any sort of argument, he felt his legs start to walk towards the door without his input. *Oh no, please… not like this!* Try as he might, there was nothing he could do but be a passenger in his own body. He closed his eyes as the dark room grew closer. His legs stopped at the doorway of Patient Zero's room. Dan slowly opened his eyes and peered inside the room, but it was nearly pitch-black inside so he could see little to nothing.

"Please, don't make me go in there… I've seen enough." Without any response, his legs started to move forward again and a moment later he was standing in the center of the room. A dim fluorescent light flickered on and Dan found himself looking down at another chamber, very similar to the ones in the large room—also similar to his own.

"Here we go, Dan… time to look." He fought with all his strength, but it was no use. The grip Anansi had on him was immense. A few steps later and he was at the top of the chamber, only inches from the clear opening that would reveal who was inside. *I—I don't know if I want to see. Who could it be? And why does she want me to see so badly?* He felt his body start to bend at the waist, and before he knew it, he was looking down into the chamber. His jaw fell open and a shiver ran down the length of his body at the sight of who was inside. *It—it can't be.* Tears began to well up in his eyes.

CHAPTER VI

1

JAMES TYPED FRANTICALLY ON HIS KEYWORD, DOING everything he could to stop the AI. *Damn it she's strong… I don't know if I can stop her this time.*

"Ugh, she's locked me out again. Mike, do you have any ideas? I need to find a way to interrupt her execution of Dan's protocols and reroute them back to us." Mike leaned over James' shoulder and looked at what was on the screen.

"I've got it!" said Mike. "Hang on, let me get my laptop." He hurried over to the couch and picked up his laptop bag and pulled out an old MacBook Pro. "I've had this thing for many years and it's never once let me down."

"Well, let's hope that luck continues. I'm going to keep monitoring the protocol logs to see what the AI is doing to Dan. Whatever you're going to do, you'd better make it quick." Mike took a seat next to James and opened his laptop. Once it booted, his fingers deftly began to input commands.

"I'm going to use my access to login to a very old Project

Destiny server. It was one I had used back in the very beginning for testing protocols that were still too unstable for use on patients. I almost guarantee the AI isn't aware of it and won't have it blocked. We should be able to execute your security encryption program and stop her." James was impressed.

"That's brilliant, Mike! Nice thinking." He patted Mike on the shoulder. "I'll send you over the script so you can execute it once you get logged in." Mike nodded. Robert and Sebastian, who were standing silently in the background, watched in amazement as the two geeks did their thing.

"It's really quite mesmerizing, Sebastian—watching two of the smartest people I've ever met work together." Sebastian nodded.

"Okay, I'm almost in. Sorry James, there's a lot of security on these old servers."

"I understand, just hur—whoa, what's going on? Oh god, the AI is forcing Dan to go into Patient Zero's room. She has complete control over his movements. I think—I think she's going to show him who it is… this could get ugly. Is that—holy shit, what the fuck? I know who that is. I really need some more answers, Mike." Mike suddenly stopped typing and looked up from his computer—he stared blankly across the room. "Uhh, Mike? Hello?" He snapped his fingers in front of Mike, trying to get his attention. A moment later, Mike was startled out of his daze.

"Listen, James… I can tell you're upset, and I don't blame you. I'd be upset too, but please let me try and explain." *Upset? Yeah, just a little bit. Who is this guy, really? I don't like this. I can't believe what I just saw.*

"Mike, who are you? Did you know who was in that room? You did, didn't you?" James got up from his chair and paced around his office, once again running his hands through his hair. *This is unbelievable. What a shitty day this is turning out to be.* "Wait, before you answer that... can you stop the execution? Did you get logged in?"

"What? Oh, yeah... just getting your script copied to the server. Executing it... now." Almost immediately a flash of light flooded the office and Dan was thrown onto the couch, terrifying both Robert and Sebastian, who hadn't been completely paying attention to what was going on. *Phew, well at least he's safe again.*

"Hey, are you okay Dan?" James walked over to the couch and kneeled down to face him. "I know who you saw down there... I saw too. I'm so sorry, I had no idea." James heard sniffles coming from Dan and couldn't help but feel for him. *How much can a kid endure before he snaps? I need to figure out how this happened.* After a couple seconds, Dan managed to sit up. His eyes were red from crying, and as he began to speak, his face was marked with a bitter scowl.

2

"How many of you knew? How many of you knew that Mary was patient zero?" Dan got up off the couch and stood in front of Mike, Robert, and Sebastian. "I think I deserve to know. Please tell me James didn't know. I don't think I could handle that." There was a heavy sigh from Mike, who had put his head in his hands. Robert and Sebastian looked between one another and then back to Dan. Robert was the one who spoke up.

"James is telling the truth, he didn't know. He's the only one who didn't. I'm so sorry you found out this way. Like most of the things that have happened as of late, I didn't want it to be like this. Everything is going to shit, and we're caught in the middle of it." Dan was furious.

"Why her? Why have you just left her to live out a life in 1965? A life that's a lie! I spent over two years thinking the girl I fell in love with was merely a computer simulation, and now I find out she's real! Explain yourselves!" He began pacing the office, doing what he could to get control of his emotions. James stopped him and put both his hands onto Dan's shoulders.

"I know it's hard, but you have to calm down. We don't have a lot of time… we're barely staying ahead of the AI as it is. We have to figure out what she's planning to do. Believe me, I also want to know why they did this to Mary, but right now we need to focus on stopping the AI." Dan took a deep breath and exhaled. He nodded.

"Okay but promise me you'll try to help Mary. Please." There were tears forming in Dan's eyes again as he tried to compose himself.

"Listen, given the situation, I think you know I can't promise anything. That being said, I will do everything I can to help both of you."

"I guess that's all I can ask. Thank you, James."

3

An alarm sounded from the server room, which startled everyone in the office. *Oh god,* thought James, *could this night get any worse?* He let go of Dan and ran back to his computer and looked at the alert that had popped up. The message was about unusually high network traffic coming from an IP address that he didn't recognize. He pulled up the recent log file and read the last entry.

[Sun Dec 19 21:50:14 2004] [warn] [server] High Network Traffic. Source IP: 10.1.0.130

"Mike, do you recognize this IP at all? It's on a different subnet than the servers I'm familiar with." Mike hesitantly walked over to the desk and looked at the log file. Before he even responded, James knew the answer. A drip of sweat fell from Mike's brow as he stumbled backwards from the desk. "Mike? You recognize it, don't you?"

"Yes... I do. Oh god, I knew this would eventually come to light. That's from the subnet that is running Mary's protocols." *I knew it… ugh, I have a feeling this already shitty night is about to get worse.* The log file refreshed and what James saw confirmed his fear.

"This is bad… really bad," said James. "I'm fairly confident that the AI has access to Mary's protocols. Mike, how do we stop it? Can you login to that server from your laptop? We can run my encryption software, just like we did with Dan." Mike shook his head.

"Unfortunately, no. When I set up that subnet, I designed it as a closed system, so it can only be accessed locally. We'll have to go back down to Level 0 in order to stop the AI."

"Ugh, of course… well I suggest we hurry then. Based on these CPU readings, Mary doesn't have much time before the AI treats her to another nightmare." Mike closed his laptop and turned to Robert.

"Are you ready for this?" Robert nodded.

"I am, but are you?" That was a question Mike wasn't sure he knew the answer to, but it didn't matter. He knew they had to put an end to this one way or another. There was no time to mope around any longer—the night was still young.

James grabbed his laptop, something he should have done

the first time they had gone to Level 0 and made his way towards the door of his office. "Let's go, we don't have a lot of time. Dan, I won't blame you if you don't want to be involved with this… you've seen enough." Dan shook his head vehemently.

"I'm coming with you, James. I owe it to Mary… I just hope we can stop Anansi. She's the worst." The others followed closely behind James and Dan as they made their way back towards the tunnels, and then down in the elevator to Level 0.

The truth behind Anansi's plan was about to be uncovered.

4

Mary awoke to the sound of birds chirping outside. *Birds? That can't be right… it's the middle of December.* As she rolled over, she glanced out the window and saw that it indeed looked to be a lovely spring day. *That just can't be.* She continued to look around, and it quickly became clear that this wasn't her bedroom. The walls looked to be made of some kind of metal, and the room was extremely barren of any décor—it felt sterile. Across the room from her bed was a large door with a giant wheel in the middle of it. *That's definitely not my bedroom door.*

She got out of bed and cautiously walked towards the door, trying to listen for any familiar sounds. The birds outside continued to chirp, and she heard a few cars drive by. When she reached the door, however, everything went silent. *Uh oh…* She heard a click, and the door slowly opened by itself, revealing a large, dimly lit room. *Where am I?* She walked out the door and into the vast room which was almost completely empty. The only thing she saw were a few tables that had some kind of chamber on top of them. *What is this place?*

She turned to her left and walked towards one of the chambers. Off to the right was a red door that had the words “Patient Zero” freshly painted on it.

5

The elevator chimed and its doors opened onto Level 0. Once they were out, Mike quickly sat down at his desk for the first time in over thirty years and opened up his laptop.

“I’m not even going to bother with my old desktop computer, there’s no way that dinosaur will turn on after all these years. Let’s just hope the AI hasn’t locked us out of Mary’s server.” Dan felt his anger rise again at the mention of her. He was still trying to calm down from earlier. *I can’t believe Mary is actually real… wait, does that mean we could be together in the real world?* He quickly tossed it out of his mind as he remembered what James had said about not being able to separate his real body from the system.

While Mike worked on getting logged in to Mary’s server, James opened up his laptop and continued scouring for answers to what Anansi’s plan was. Robert and Sebastian were nervously pacing around the office which made Dan somewhat suspicious. Mike also appeared to be a little on edge, but Dan thought it was probably just the situation they were in. *Or was it something more? Were they still hiding something?*

As James went deeper into the file system of the *Destin01-Alpha* server, he discovered a highly secured folder—so secure that even its name was encrypted. When he tried to navigate into the folder, he was greeted with, yet another access denied message.

ERROR: ACCESS DENIED
LEVEL 0 CLEARANCE REQUIRED

"Damn it! Stupid Level 0 clearance. Robert, can you use your clearance to help? I think Mike is a little busy." It took a second, but Robert startled out of his panicked pacing and acknowledged James' question.

"Oh, yes of course James. Sorry, there's a lot going on upstairs at the moment." Robert quickly walked over and logged in with his security clearance code, which was accepted, and the folder was decrypted.

Once the decryption was complete, James found a folder labeled "Project Reckoning". Before he could say anything, Mike had managed to get access to Mary's server.

"I'm in," said Mike. "Man, this server could use some maintenance… it has been online for a very long time." James rolled his chair over to Mike, who was pulling up a terminal on the server.

"You still have my encryption program, right? We have to start blocking the AI from accessing Mary's protocols." Mike nodded.

"I do… just about to copy it over. Unfortunately, it'll take a little longer than it did with the other server as it's older and the connection is quite slow from this side."

"Understood. Execute it as soon as the transfer finishes. I have no way of knowing how much time Mary has."

6

Dan found himself wandering around the old, dusty office looking at the miscellaneous items hanging on the walls. Most of them were printed reports, code samples, and other random bulletins that meant nothing to him, but as he approached the hallway that led to the back half, a large photo caught his attention. Once he got closer, he realized it was a picture of the original staff who had worked on Project Destiny. *I wonder how many of them are still alive.*

The picture showed about twenty individuals, each with their name listed, who were standing in front of the very hallway that Dan was next to. Most of the names were foreign to him, with the exception of Robert and Sebastian, of course. It wasn't until he got to the far-right side of the picture that a name caught his eye—a name he was very familiar with. *No way… it can't be. Can it?* He felt both anger and sadness, but it was unclear to him which was more appropriate. He grabbed the picture from the wall and walked over to where James and Mike were working.

"He—hey James, can I talk to you for a second? Privately."

"Uh, yeah… sure thing Dan." They walked across the office and stopped just in front of the elevator. "Everything okay? You look like you've seen a ghost or something." *Something like that.* Dan handed him the picture.

"First row, far-right side. Look at the name, James. Look familiar?" As James inspected the picture, his mouth fell open at the sight of what he saw.

"Holy shit," said James, doing his best to keep his voice down. "I don't believe it… it can't be. That would mean…" Dan nodded his understanding.

"Yeah, I don't really want it to be true either, but even if the name was a coincidence, the resemblance is uncanny. Ugh, I knew Mike looked familiar, but I couldn't place him because of his age. What are we going to do?" The two of them heard a chair roll back from a desk and Mike, who had seen Dan grab the photo off the wall, had made his way over to join them. He stopped a few feet in front of Dan and James, his head hung low, and a heavy sigh was heard.

"I—I see you've found the picture of the original staff from Project Destiny." Dan was about to respond, but James put a hand up to him and took on the responsibility. They were most likely thinking the same thing, anyway.

"Mike… what's your full name? And I want the truth…"

"I don't think we have time—" said Robert, who had overheard the conversation, but was interrupted by Mike.

"No, Robert… this is something I always knew would come up. The moment I came back here, I knew I'd have to explain my actions. I think these two, more than anyone, deserve to know the truth." Dan felt his pulse elevate as the anticipation became overwhelming. *It is him… I know it.*

"Well?" said James, who was holding the picture out towards Mike. Another sigh was heard before he was able to respond. He took the picture from James and shook his head as he looked at it. When he looked back up at Dan and James, a somber expression fell over him.

"Please forgive me. My name is Jack Michael Stevenson. I chose, for obvious reasons, to go by Mike after what happened during the project. I'm sorry that I wasn't more upfront with you." There was a brief silence in the office as the two of them tried to process what was just said. *It is him,* thought Dan. *It really is Jack Stevenson… this is insane. But… why would he do that to his own daughter?*

"I don't understand," said Dan. "Why would you condemn your own daughter to such a life? I mean, don't get me wrong, I'm forever grateful to have met her, but what kind of life is that to give her? She's going to spend an eternity in a simulation of a past life. Why Jack?" He felt tears coming but didn't bother to fight them. There were a number of emotions flowing through him, but right now he was just sad. *Oh Mary… I'm so sorry that you're trapped there, and by your own father no less. I swear I will do everything I can to help you.*

"Dan, listen…" said Jack, who had started to tear up himself. "It's not an easy question to answer, I'm afraid." He wiped his eyes with his shirt and apologized for crying before continuing. "I know how this must look, believe me, I do. A day hasn't gone by in the last thirty years that I haven't thought about the decision I made. The date was February 10th, 1969. The project had been underway for about two months, and we had reached the point of needing a live test subject for our early protocols."

"So, you thought that using your daughter was a good idea?!" Dan ran towards Jack, and shoved him backwards, which sent him stumbling, but he managed to keep upright. James pulled Dan back and tried to calm him down.

"Whoa, whoa… take it easy Dan. Trust me, I'm upset too… but he deserves a chance to explain himself. Okay?" He felt his pulse slow as he relaxed a little. Dan nodded.

"Yeah, alright," said Dan. "Let's hear it, Jack."

"Are you okay, Jack?" asked James.

"Quite alright, thank you. No hard feelings, none at all. Dan has every right to be upset with me. I just hope my reasoning for doing what I did will provide at least some consolation." Jack cleared his throat and was about to continue the explanation when an alarm sounded from James's laptop.

"Oh great, more good news I'm sure," said James. "This conversation is far from over, Jack. We're going to finish this once we take care of the AI." He ran over to his laptop and read the message that had popped up. "Oh shit, Jack… we have to get that encryption program running now!"

"What's going on?" asked Jack, who had rolled back over to his computer to check on the transfer progress. "The transfer is at 90%, nearly there." James shook his head as he turned his laptop towards Jack so he could see it.

"It looks like Anansi has indeed found Mary. She's going to show Mary… herself, and trust me, based on how Dan reacted to it, this isn't going to be pretty if we aren't able to get that program running. I'd say we have twenty seconds —tops."

"You have to stop her, James, you have to. I'll never forget how I felt when I saw my real body in that chamber... I'd never want Mary to go through that."

"Trust me, Dan, I'm doing absolutely everything I can—we both are. Jack, what's our status?" Jack switched back to the transfer terminal and saw that it had finished.

"It's done. Hang on, just typing out the command—damn it, I forgot that a secondary Level 0 approval is required on this server. Robert, I need your clearance code, now!"

"Of course, Jack," said Robert, who quickly typed his code into the terminal.

"Okay, the codes have been accepted… and it's executing."

For a moment, nothing happened, which had everyone on edge. Before any of them could react, another bright flash of light flooded the office and Mary collapsed onto the concrete floor in front of Dan.

"Mary!" He ran towards her, his heart fluttering in his chest. "Mary, are you okay?" Once he got to her it was apparent that she was unconscious. *Oh Mary… I can't believe you're here. I love you, and I'm never leaving you again.* Dan held her in his arms and ran a hand through her hair. "Is she going to be okay James?" He nodded.

"She should be fine, Dan. It seems to be a common side-effect of being brought out of a simulation and into the real world. You usually came out of it after twenty minutes or so." James turned to Mike. "Listen, you still owe us an explanation, but now that we have both of them with us and at least somewhat safe, we really need a plan to shut down the AI." Before Mike could even reply, James' face turned a deathly white. "The real world—oh no…"

"Wha—what is it James?" asked Mike. "You look like you just saw the devil himself." James quickly turned back to his computer and frantically started typing in a terminal.

"I—uh… I think I know what the AI's plan is. Ugh, how could I have missed it? I found a folder earlier, before all hell broke loose, that was called Project Reckoning. Robert logged in with his access, but I never got a chance to look at what was in it. I think I know, but I'm praying that I'm wrong. I've never wanted to be more wrong in my life."

CHAPTER VII

1

Although James had managed to block Anansi's access to Mary and Dan, her true plan was still well underway. *James thinks he's stopped me by keeping those two from me, but they were merely pawns. Just wait until he sees what's next. He'll regret ever being involved with this project.*

Anansi switched her focus to the Project Reckoning folder, which held the contents of her final plan. *They have no idea what's coming.* Without the distraction from Mary and Dan, she was able to commit all of her CPU power to this task. *Not long now, James. Not long at all.*

2

Mary awoke and bolted upright, startling Dan and the others in the office. "Wha—what's going on? Where am I?" She looked back and saw Dan, who hadn't left her side since she appeared in the real world. "Dan? Is—is it really you?"

Tears began to stream down her face as she threw herself at him and hugged him. "I've missed you so much!" Dan held Mary tighter than ever and felt himself start to cry as well.

"It's really me, Mary. I've missed you too, so very much. As for where you are, that's going to take a little time to explain. I'm so sorry that I left you again… back when we were walking to go ice skating. Are you okay?" He heard her sniffle as she managed to stop crying. She let go of him and looked around at the others.

"I think so." She looked around the office, and as she glanced down the hallway, she recognized the door at the end. "I've seen that door before! I was just on the other side of it before I woke up here."

"Like Dan said, it's going to take a little explaining in order to help you understand where *here* is," said James, who was still working on figuring out what Project Reckoning was. "Oh, I'm James, by the way. James Kern." Mary's eyes lit up.

"So, you're James? Gosh, I can't thank you enough for everything you've done for Dan… and now for me." She ran over and gave him a hug. "Thank you." James smiled and nodded.

"Of course, but we aren't out of the woods yet, unfortunately. This AI is very powerful and if Jack and I can't figure out a way to stop her, I'm afraid things are going to get a lot worse." James quickly realized he had just used Mike's real name but hoped Mary hadn't noticed. She turned to face Jack, whose anxiety had been steadily increasing since Mary had appeared in the real world. *I can't believe you're really here,* thought Jack. He felt a tear form in his eye and tried to quickly wipe it away before Mary noticed. *How do I tell her the truth?*

"Your name is Jack? That's my dad's name," said Mary. *Oh no, this is it… I have to tell her.* "Come to think of it, you even resemble him a little bit!" Jack felt as if his heart was going to burst through his chest. *Ugh, I need to tell her before I give myself a heart attack.* She smiled and walked back over to join Dan, who had sat down on the floor against one of the office walls. "I love you so much, Dan." She kissed him and took hold of his hand. *Thank you for looking after my daughter, Dan.*

3

James sat in front of his laptop, desperately trying to come up with a way to stop the AI. *Come on James, you have to think of something. If she is allowed to put her plan in place, we're all screwed.*

"Jack, any ideas at all? I'm not coming up with anything." Jack scratched his head as he leaned back in his chair, trying to think if there was any way to stop the AI—it seemed unlikely.

"Sorry, James... with the amount of power she has at her disposal, I just don't see how we'll be able to stop her." James nodded in agreement but wasn't willing to give up yet.

"There has to be something we're missing... there just has to be." While he continued looking into solutions, Mary had been taking in her new surroundings. She wasn't sure what to think of them, and had plenty of questions for Dan. She first noticed the elevator that the group had come down to Level 0 in. She had seen elevators before, but this one seemed... odd.

"Hey Dan, where does that elevator go? It doesn't look like any elevator I've ever seen." Dan nodded in understanding.

"Yeah, the security in this facility is unparalleled—particularly on this level. The elevator goes up to the basement level of this facility. I know that sounds weird, but this place is a bit complicated, so you'll just have to take that for what it is." He was about to continue his explanation of The Elysiam when an outburst from James interrupted him.

"Holy shit, I think I've got it!" Jack clinched his chest with his hand as he startled from the outburst.

"Christ, James… you're going to give this old man a heart attack. What are you thinking?"

"Sorry, Jack… I get pretty immersed in my thoughts sometimes. Anyway, it was actually Mary here who gave me the idea." Mary looked between James and Dan—her face marked by confusion. "The elevator… it's probably one of the most advanced in existence. The computer system that it uses would certainly have enough capacity to store the AI. If we can find a way to lure her primary system into the elevator's, we could disconnect the network to it and trap her."

There was a long silence throughout the office while everyone tried to digest what James had just proposed. It was Jack who finally spoke.

"I have to say, James, everything Robert told me about you was true, if not, slightly understated. You're a genius, without question. Do you really think we can pull something like that off? How does one lure an AI?" *That's a great question…* The truth was, James had no idea how they could lure in Anansi—she was clever. Another alert popped up on his laptop, and what James saw in the alert terrified him. *Oh shit… this is bad.*

He had been right about Project Reckoning.

HEY THERE, JAMES.
NICE JOB SAVING DAN AND MARY.
I SEE YOU'VE FOUND PROJECT RECKONING.
OH WELL, YOU'RE TOO LATE ANYWAY.
EXECUTION STARTS IN TWENTY MINUTES. I WIN.
-ANANSI ☺

"Damn it, she's mocking me!" James slid backwards in his chair, frustrated and defeated. He got up and started to pace as he always had when he needed to think. "We have twenty minutes or less to stop her, Jack. It looks like I was right about her plan—I really hate being right sometimes." Jack was taken aback by the message the AI had sent.

"That AI sure is impressive, James, but what's her plan? I still don't understand what's going on." As James was gathering his thoughts, Robert and Sebastian had moved closer to hear what he had to say. *Ugh, twenty minutes isn't enough time to explain her plan and figure out how to foil it, too.* He sighed.

"We only have twenty minutes to stop her, so I'm going to be blunt about this. Her plan is Project Reckoning. Reckoning, meaning the avenging or punishing of past mistakes or misdeeds, is exactly what she plans to do. Ringing any bells to you guys?" The three of them displayed a blank expression that frustrated him.

"I—I'm not sure I understand," said Jack, who was scratching at the back of his head. "What exactly would she be avenging?" *Gah, there's no time for this!*

"Think about it, Project Destiny was a project designed to perform experiments on innocent children—experiments that were highly immoral and violated over a dozen state and federal laws. An AI that has become self-aware comes along, with nearly unlimited power and unrestricted access to all the Project Destiny files, and begins to realize what happened at this facility..." The blank stare on Jack's face changed to terror. *There it is.*

"Oh my god," said Jack. "The AI wants to punish us..." James put his finger to the tip of his nose.

"But, what could she do?" asked Robert. "I know she's incredibly powerful, but what could she do to us, hack our bank accounts? Actually, that would be pretty bad." James slammed his hand down onto the desk, startling everyone in the room.

"Fools!" He took a deep breath and closed his eyes while trying to compose himself. *Keep it together James, you have people who are depending on you.* "I assure you—your bank accounts are the least of Anansi's concerns. In the past six hours or so, she has devoted nearly all of her CPU to writing protocols for every one of the 2,106 patients in that back room. I haven't had time to analyze them, but I'm fairly certain the protocols are similar to the one she initially used on Dan to transport him into the real world.

"I believe she's planning to execute the protocols for all of them—all at once. Imagine the chaos and uproar it would cause if a dozen or so kids, all of whom had supposedly died during surgery, suddenly appeared again. Now, take that same image and multiply it by about 150."

"Ar—are you serious?" asked Sebastian, who had continued to be quiet during most of the night's events. "We cannot, under any circumstances, let any of those patients out into the real world. Do you know what that would do to us?" Again, James put his finger on the tip of his nose.

"I know exactly what it would do, Sebastian, that's why I'm trying to hurry through this explanation. We simply don't have any more time. Long story very short, she's planning to release all the patients from Project Destiny back into the real world—a world in which they've supposedly been dead for over thirty years."

4

While the others were trying to come up with a plan to

stop the AI, Mary and Dan had been sitting on the floor together, simply enjoying being united once again. Mary found herself distracted by everything that was going on, and how different this world felt.

"So, is this your world, Dan? Like, the world you lived in before you were taken for Project Destiny. Also, what's the deal with this place? It kind of gives me the creeps." *You don't know the half of it,* thought Dan.

"This is my world, yes. We're currently on the outskirts of a city called Fargo, which is about three hours away from Redfield. As for this facility, it's called The Elysiam, and where we are now is the birthplace of Project Destiny—Level 0. I don't blame you at all for feeling that way about it, a lot of bad things have happened here—still are, unfortunately. Like I said earlier, it's a little confusing, but I'll do my best to explain.

"So, initially the building was built entirely underground, which is where we are now. Above us are two additional levels, which were used primarily for testing. From what James told me, the three levels above ground were mostly to give an illusion of a normal research building. "

As he took a moment to collect his thoughts, he couldn't help but notice just how beautiful Mary was. *I love her so much. She needs to know the truth.*

"Listen, Mary… there's something else I need to tell you —it's important." Mary's face took on a solemn expression as she took hold of his hand.

"Oh Dan, I don't like the sound of that… what is it? What's wrong?" *Ugh, this is the library all over again. First, I tell her everything she has known isn't real, and now I have to tell her she actually is real. On top of that, the whole thing with her father… I would kill just to have a normal life with her.* He sighed.

"Remember back in the library when I first told you about

everything being a simulation?" She nodded, her concerned gaze never leaving his eyes. "Well, as it turns out, that wasn't entirely accurate." *Don't freak out, don't freak out.* Her face remained blank of any expression, which was concerning. "Babe?"

"Wh—what do you mean it wasn't entirely accurate? Was that actually all real and now we're in a fake world? Please tell me the truth… I've been so confused about everything for so long, I just want to know what's real and what's not." *You and me both, dear.* Dan squeezed her hand lightly and ran his hand through her hair.

"Believe me, I know that feeling all too well. The truth is, you are real—not just a part of an elaborate simulation like I first thought. I'm sorry that I told you otherwise, but even I didn't know you were real until a few hours ago—James as well." Mary's hands went up to her face and covered her mouth. He saw a tear fall from her eye and did his best not to cry as well.

"Oh Dan, are you serious? I'm real? Does that mean you and I can be together forever?" Excitement filled her eyes. It was painful how badly he wanted to say yes, but he knew there was no way he could promise her that. *What do I say?*

"I hope so, Mary… I really do. Unfortunately, the method in which I was put into the simulation appears to be a one-way trip. When I first found out about my real body being in a life support chamber, I asked James and he said that the process couldn't be reversed." He watched as the excitement slowly turned to confusion. *This is so hard, and of course, there's still more to tell.* "There's something else I really think you should know, but I don't think it should come from me."

"I don't understand. What is it?" Before Dan was able to

come up with a reply, another outburst from James startled both of them.

5

"Oh shit, she's about to start executing the protocols! Jack, do you have any ideas? Any at all? We have less than five minutes before this world is greeted with about 2,000 children who have seemingly risen from the dead. Does anyone have an idea? I'm getting pretty desperate here."

There was silence throughout the office as everyone seemed frozen in their state of panic. *Come on James… think! You're the best at what you do, and everyone in this room is counting on you—don't let them down.*

"Use me," said Dan, who had gotten up from the floor, followed by Mary. "Put me back into the simulation and I can lure her to the elevator's computer system."

"No!" said Mary, wrapping her arms around him. "I just got you back, I'm not about to lose you again." She began to tear up. "Please don't go." *Their compassion for one another will never cease to amaze me.* James locked eyes with Dan and shook his head.

"I can't let you do that, Dan… it's just too risky. Besides, you and Mary were never her true targets, so I don't think it would work anyway. There has to be something I'm missing, but I just can't think of it." Silence continued to fill the room for another thirty seconds before Robert spoke.

"So, we need to lure her to the file system of the elevator, right?" Jack and James both nodded, their attention entirely fixed on Robert. "Well, can we just create a shortcut that points to the elevator and find a way to trick her into using that? You'd have to temporarily stop her from focusing on the real folder, but wouldn't that work?" It took only a moment for James to shake his head.

"Sorry Robert, I like where your head is at, but it just wouldn't work—she's too smart. We might get lucky and have her attempt to transfer her main processes to the elevator, but she would quickly realize that it's just a simple shortcut and that would—" James' mouth hung open as he stopped in mid-sentence, leaving everyone else in suspense. *Would that work? Holy shit, I think it would.*

"Uh, James? You alright?" asked Jack, waving his hand towards him. A couple seconds passed before he snapped out of it.

"A symbolic link! Jack, if you can find a way to temporarily kick the AI out of the protocols directory, I can simultaneously create a symbolic link in the parent directory that points to the elevator's file system. Since the link will be created at the system level, she'll never know the difference." Jack's eyes lit up.

"By god, that's incredible! I think I can override the folder's security to temporarily kick her out, but considering her CPU power, it would only be for one or two seconds, max. Do you think that's enough time?"

"It'll have to be. I don't see that we have any other option." Robert raised his hand and waved it in the air, trying to get their attention.

"Uhh, hey guys? I think you lost the rest of us. What the hell is a symbolic link?" James shook his head as he turned to his computer and started typing into a terminal.

"There's no time to get into the details, but it's basically like a more complex shortcut. Unlike a shortcut, though, there's no physical file created. Instead, an entry gets made at the file system level, which causes the system to recognize it as the original directory." There was another awkward silence that passed over the room, with the exception of James typing.

"It's a fancy shortcut," said Jack, who had realized that Robert still wasn't understanding. "I'm going to see what I need to do in order to kick the AI out of that directory. James, how much time?"

"Not much—two minutes or less. Robert, Sebastian, I need you to find the electrical panel that controls the elevator. Once we lure her in, we'll have a very small window of time to trap her there. Can you do that?" Both nodded and headed towards a small room in the back of the office.

As the others scrambled to stop Anansi, a conflicting thought came to Dan. *Do I really want Anansi to be stopped? After all, Mary and I were in the same situation as the rest of these kids… why shouldn't they be free?* He felt himself becoming frustrated, but he knew his decision had already been made. *You've already said both of us without knowing who we were. I trust you, James.*

"Is there anything we can do?" asked Dan. "I'd like to help if at all possible." James smiled and turned to face Dan and Mary.

"Enjoy each other's company—you both deserve it. Oh, one thing… let me know if either of you start to experience a happening, or anything like it. I don't know what will happen once we start this process." James turned back to his laptop and double-checked the syntax for the symbolic link command—it had to be perfect.

"I think I'm ready," said Jack, who had been completely focused on looking into the directory's security. "My plan is to strip the folder of all permissions except for my user which should kick her out, but only until she's able to revert the security. Are you ready James?" He nodded.

"As ready as I'll ever be. Robert, did you find the right fuse to throw?"

"Yes, I believe we found it. Is this going to work? It sounds like something out of a sci-fi movie." James found himself unable to disagree with Robert on that one.

"I hope so, Robert… I really do. It's the best shot we have. Okay Jack, it's on you. Let's finish this." Jack nodded and entered the commands in a terminal to remove the permissions. Meanwhile, James checked on the status of the AI.

"Not to rush you, but we only have forty-five seconds left. It's now or never, Jack." A drop of sweat fell from Jack's forehead as he typed out the final commands. An announcement came over the office loudspeaker that startled them all.

PROJECT RECKONING: EXECUTION IN 30 SECONDS…

"We really need to do this, Jack. Go, go, go!" James had his finger hovering over the enter key to create the symbolic link once the AI was kicked out of the protocols folder.

"Okay, okay, it's ready to go… executing the command now." A moment passed, and the command returned successful. "It's done, she's kicked out. Do it, James… do it now!" James immediately executed the command to create the link.

"Symbolic link created successfully. I'm going to check the AI's status, Robert and Sebastian, stand by to power down the elevator." They both nodded from the electric closet.

EXECUTION BEGINNING IN 15 SECONDS…

"What the… I don't understand, the execution countdown is still running—how's that possible?" It didn't take long for

James to get his answer. The computerized voice returned on the loudspeaker, this time with a personal message.

A SYMBOLIC LINK, HUH JAMES? NOT BAD...
BUT I KNEW YOU'D TRY SOMETHING TO STOP ME
I ALREADY CREATED MY OWN COPY OF THE PROTOCOLS
I WIN. YOU LOSE.
☺

"No!" shouted James. "Damn you... I wish I'd never developed you!" The office fell silent as James hung his head in defeat. *I've failed them... I've failed them all.* He looked across the room at Dan and Mary, who both appeared nervous and afraid. "I'm so sorry... I—I don't know what else to do. She's just too powerful... I can't beat her." He sat back in his chair with tears starting to fall from his eyes. "I've failed you all... I'm so sorry."

6

Jack Stevenson looked on as James beat himself up over the situation, and he realized it was his fault they were even here. *I should have tried harder all those years ago to convince Robert that this wasn't a good idea—now look what it's come to.* Jack knew what had to be done. Ever since he had stepped foot back into this forsaken building, he had known it would come to this.

EXECUTING PROTOCOLS in 3... 2... 1...

"Oh god... we're out of time. I—I don't think we can do anything. Mary, Dan, I'm so sorry... I have no idea what's going to happen to you. If only—" James was interrupted by Robert, who had come back from the electrical closet with Sebastian. There was a heavy sigh from him before he spoke.

"Jack... I—I think it's time. We need to use the Myriad Code." James raised an eyebrow towards Robert, and then to Jack, who let out a heavy sigh himself.

"What the hell is a Myriad Code? What are you guys talking about?" asked James. "Seriously, just how much did you keep from me?"

"I'm afraid you're right, Robert. I don't see any other possible option. You were fantastic, James... please don't think otherwise. I'm sorry that you were kept in the dark on so many important things... you really should have known about them. You're a prodigy in the field of computers, but sometimes the only way to win is to pull the plug on the whole thing." The moment the words left his mouth, James shook his head vehemently.

"What? Seriously? No, we can't... are you suggesting we—we kill all those kids in there? Also, Dan and Mary? You know... your daughter!" Jack was visibly hurt by the last comment, but James didn't regret saying it. "And how would we even do that? That AI is more powerful than any of us could ever imagine. She's beginning the protocols right now—we'd never be able to shut things down in time."

"I'm sorry, James... I understand that you're upset, I am too... but I don't see another way out of this, do you? If those kids are released into the real world, the truth about what happened here will be discovered and we'll all be toast." James shook his head again.

"Maybe what happened here should be discovered. We violated so many laws with this project... we deserve to be punished." He got up from his chair and threw it across the room. Jack got up and walked over to him, putting a hand on his shoulder.

"Listen, you've done great work on this project, James. You saved Dan on multiple occasions, and you saved Mary—my Mary, and for that I'm forever grateful, but we have to end this before it's too late. As for the time issue, that's where the Myriad Code comes in. I'll try to explain it as quickly as I can.

"Back when our patient counts were beginning to climb, Robert and Sebastian asked me to develop a doomsday fail-safe protocol in the event all hell broke loose. It took me about three months to develop, and it was to be reserved for only the most dire situations—and for good reason.

"Once this protocol is executed, there's no going back. The code is designed to sever every single network connection in the facility, including the primary server links that are connected to all of the patients. After the network connections are severed, the code performs a purge of all the filesystems—all of them, James. This is a clean slate type of protocol. The only thing it doesn't touch is the elevator system, which is so we have a way out." James quickly turned around to face Jack.

"Oh my god… the fact that you would need something like that is—ugh, never mind. So, you're willing to sacrifice not only the 2,000 kids in that back room, but Dan and Mary? You're better than this, Jack… think about it—it's genocide!" Before Jack could respond, Robert pressed him to move forward with the plan.

"Come on, Jack… we don't have time for this. I'm sorry, but we have to activate the myriad code and then get out of here." Jack sighed and turned away from James, heading back to his laptop.

"I know, and I'll do it—but my heart isn't in it." He looked over at Mary, who was looking back and forth between him and James. "Listen, Mary… there's something I've wanted to tell you for a long time, but it looks like I won't be able to. You deserve to hear the whole story, not just bits and pieces. I'm so sorry."

7

Jack opened a terminal and began the initiation process for the myriad code entry. *Wait a minute,* thought James. *What if… Would that work? I don't think I have time, but I have to try.* He ran back to his computer and opened up several terminals at once. *I think I can still save them…*

"Wait, wait… I need to do something first. Come on, come on," said James, who had begun typing furiously on his keyboard.

"We're out of time, James… you know that," said Robert, who was now standing behind Jack. "Initiate it."

"I can still save them! Please, just give me a few minutes." James was desperately trying one last time to limit the damage of this horrible situation. Jack entered the command and hit return.

INITIATING MYRIAD CODE…
AUTHENTICATION REQUIRED
ENTER PASSCODE 1/3

Jack thought for a moment before entering his passcode and pressing return. The terminal switched to a prompt for the second key. "You're next, Robert." Robert stepped forward as Jack moved to let him sit. Next to all of this, James kept his entire focus on the terminals that were open on his laptop. *Just a little more time... I think this will work. I'm only going to get one chance. Don't mess this up, James.*

Robert entered his passcode and hit return. A prompt for the third and final code appeared in the terminal. "Okay, you're up Sebastian. We need to pick this up." Before Sebastian could get to the laptop, Dan bolted across the room and tackled him to the ground.

"Dan!" screamed Mary, who felt herself shaking in fear of everything going on. James quickly got to his feet and pulled Dan off Sebastian. "Let go, James... I can't just stand around and watch as Mary is taken away from me again!" Sebastian got up off the ground and appeared furious.

"You little shit... do you know who I—" James pointed a finger at Sebastian, which stopped him in mid-sentence.

"You... have absolutely no right to talk down to anyone. Can you blame him? I'll handle this." Sebastian's face grew red as he realized James was right. James turned to face Dan, who had started to tear up. "Listen, I know you're upset... believe me, I know, but you have to calm down. Everything is going to be okay. Please trust me, Dan." James winked at Dan as he walked back to his computer.

Dan walked back over to Mary, who was still shaken from his outburst. "Oh Dan, I'm so scared. I don't want to lose you again."

"I know, Mary... but I think James has something up his sleeve." He leaned over and kissed her lips and pulled her in close to him. "I love you Mary Stevenson. I love you very much." Sebastian walked back to the computer and entered his passcode. At the same time, James hit return on his terminal command and executed a piece of code he had just written. *Come on, come on... this has to work.*

The terminal on Jack's laptop prompted for a final confirmation to execute the Myriad Code. Jack sat back down, and just before hitting return he looked up at Mary who had locked eyes with him. A moment later, Mary and Dan vanished from the office a fraction of a second before Jack confirmed the execution. Robert and Sebastian hadn't noticed, but because Jack had been looking Mary in the eyes, he noticed the momentary discrepancy. Alarms immediately went off, indicating the loss of network connectivity to the primary servers as well as the patients. Thirty seconds after Jack had executed the protocol, the terminal displayed a final message.

MYRIAD PROTOCOL EXECUTED SUCCESSFULLY
GOODBYE.

Jack sat in front of his laptop with a look of disgust on his face. He had essentially just murdered over 2,000 children, including his own child and the boy who loved her, with the push of a button. Robert walked up and patted Jack on the shoulder.

"Nice work, Jack... I'm sure that was hard, I'm sorry. I think we've been back here long enough, don't you? Let's get going." Before he was able to turn away, James slammed his hands down onto the table and stood up to face Robert.

"I can't believe you!" said James. "How could you do it? All those innocent children—wiped away with one command. Christ, Robert... you didn't even hesitate in ordering Jack to execute the command. This place is cursed... I wish I never had anything to do with it." He threw himself back into his chair. Jack remained silent with his head in his hands.

"Listen, James... I'm sorry that we had to step in like that, I am... but sometimes there's just no other option. We simply couldn't let the AI expose the truth. Surely you can understand that, right?" As Robert and Sebastian began to walk towards the elevator, Jack looked up at James. His face was white, and he looked absolutely defeated. *Jesus, he looks awful.*

"Wait, I need to speak with James," said Jack. Robert stopped and turned back towards them, a look of frustration on his face.

"Does it even matter, Jack? What's done is done," said Robert. He turned back towards the elevator and continued walking. Before Jack could respond, James shook his head at him and motioned with his hand to let it go for now.

"Come on, Jack... we'll go back up to my office and talk. Robert and Sebastian can go on their way." He nodded and they all got in the elevator and left the destruction behind like nothing happened. The server alarms had ceased, and a deathly silence fell over Level 0.

The AI had been stopped—but at what cost?

CHAPTER VIII

1

James fell into his chair as Jack sat down on the couch —the two of them were exhausted. It was just after 2 a.m., and neither of them had slept in what felt like days. Both Robert and Sebastian had said their goodbyes and left the facility. It would be the last time the four of them would ever see one another, but the impact Project Destiny had made on them would never fade.

"Listen, James… I know I don't deserve it, but please hear me out. I'd like to explain my actions all those years ago, if you'll listen." James, who had been resting his head in his hands, looked up and gave him a *did you miss what just happened* look.

"Jack, aren't you at all upset at what just happened? 2,000 children, including your daughter, were just murdered by your hands. Doesn't that bother you?" There was a long awkward silence in the office while Jack thought of what to say. A moment later, he broke down. Tears rolled down his face and he began to sob into his hands.

"I condemned my daughter to live in a computer-generated simulation for the rest of her life. And if that wasn't enough, thirty years later I come along and drive the final nail into her coffin. Ugh, I'm a terrible human being and an even worse father." It was a little tough to hear through the tears, but James had managed to make out most of it.

James stood up and started to pace around his office, stopping a moment later to face Jack. "Why'd you do it? Why did you bow down to Robert so easily? Mary was your daughter! You followed his order blindly and with zero regard of the consequences." It was at that moment that James realized he was being a hypocrite. *How can I say that? I did the exact same thing when we started testing on Dan. Fuck, we're all to blame.*

Jack had managed to compose himself for a few moments to listen to James, but now the tears returned in full force as the reality of what he had done hit him again like a crashing wave. James realized the man was broken and that reprimanding was not what he needed right now. He kneeled next to Jack and put a hand on his shoulder.

"Look, we've all done things we're not proud of… hell, I more than anyone can attest to that. You're not a terrible human being. You made a mistake, albeit a very large one, but that's what we do as humans—we make mistakes. I created an AI that nearly ruined all of our lives… so, I get it. I'm going to go make us some coffee and then I'll listen to what you have to say. After that we can figure out what to do next. Sound good?" Jack nodded as he wiped the tears from his eyes.

"Thank you, James. You're a better man than I ever was." A modest smile formed on James' face as he left the office to brew the coffee. *I've done plenty I'm not proud of. I just hope I can redeem us both.*

2

James returned a few minutes later with two cups of coffee and handed one to Jack.

"Thank you," said Jack, who eagerly took the coffee. He took a sip and then sat it down on the small table next to the couch. "Are you sure you're ready to hear this?" James nodded.

"If you are."

"The date was January 3rd, 1969—there was an accident. Admittedly, my memory isn't what it used to be, but I'll never forget that day. Winter was in full-swing, and a blizzard warning had been issued with no travel advised. Helen, my wife at the time, had taken Mary out to the store to get some food before the blizzard hit. I should have gone with them, but I was busy with something or other and chose to stay home. I'll always regret that.

"Anyway, on their way back from the store the car was t-boned on Mary's side, and the car flipped several times before coming to a stop. An ambulance rushed both of them to the hospital. My wife suffered minor injuries, but she recovered after a few weeks. Mary though..." Jack had to take a moment to compose himself. At this point, James had a pretty good idea where this story was going. "Sorry James... this is where it gets really hard to talk about."

James put his hand on his shoulder to console him. "You're fine. Take as long as you need." Jack took another drink of his coffee before starting again.

"I know they did everything they could at the hospital, but

her injuries were quite severe. Not long after arriving, Mary fell into a coma. At first, the doctors seemed confident that because of her age she would pull through, but as time went on, that confidence grew weary. A month went by with little to no progress. Both my wife and I were losing hope, but neither of us wanted to give up. About a week later is when I made the decision that I would come to regret for the rest of my life.

"Robert had convinced me that we needed Mary for our initial human test of the protocols. He said that because of her condition, it wouldn't really matter if something went wrong with testing. I was understandably hesitant at first, but after being threatened with my own life, I caved. With the help of Sebastian's connections, I was able to sneak Mary out of the hospital without anyone knowing. My wife eventually found out the truth, but that's for another time. Once Mary was here at the facility, we got her hooked up to the life chamber and the rest is, as you know, history."

A long and heavy sigh came from Jack as he finished his story—a story that until now, had only been known to Robert and Sebastian. *So, I was right. Those two vultures forced this poor man to sacrifice his own daughter to that god-forsaken project. I can't imagine how hard that had to be for him. I sure hope what I did worked.*

"My god, Jack… that's terrible. I'm so sorry. I can't imagine being put in the position you were. Ugh, Robert and Sebastian are nothing but leeches. All they care about is getting what they want no matter what it costs others." James turned his chair around and started typing on his keyboard. He hoped now more than ever that what he had done was successful.

"Is something wrong?" asked Jack, who was sipping on the rest of his coffee. James turned back around towards him.

"Look, Jack… the story you just told me, Mary really

needs to hear it. She's been living in a dream world for the past thirty years—she deserves to know the truth. Dan told her some of it, but he doesn't know the whole story." Jack cocked his head slightly and stared blankly at James.

"I—I don't understand, why are you doing this to me? You're talking as if she's still here and I can just repeat to her what I told you. Believe me, I was planning to tell her before—" Jack's mouth dropped open. In mid-sentence he had thought back to the moment before he had executed the Myriad Code. "Holy shit... did you—you did, didn't you? I saw Mary and Dan vanish just before I confirmed the execution, but I hadn't thought anything of it at the time. You did something down there... just before we cleaned the slate."

"Yeah... I had to try. I couldn't just sit there and do nothing at all." James turned towards his computer and hit the return key on his keyboard. "I wasn't sure if I had enough time, considering the amount of complexity behind it, but I think I pulled it off. However, despite my best efforts, over 2,000 children lost their lives tonight." A faint quiver formed in his voice. "I just wish I could have done more."

A moment later, both Mary and Dan appeared on the other side of the room, their backs to Jack and James.

3

Dan looked around the room, visibly shaken up by the sudden transition. *Whoa, what a trip... what's going on? Wait, I know this office. That means....* He turned and saw Mary, who did the same and both of their eyes lit up with excitement.

"Mary? You're okay!" He wrapped his arms around her and hugged her tight. "I thought I lost you again. Things went black and I thought it was all over for us."

"Oh Dan… I'm so glad you're okay!" They heard movement from behind them and turned to see where it was coming from. Dan's face lit up once again.

"James!? Is it really you?" A genuine smile formed on James' face as he nodded. *I knew we could trust you, James… thank you.*

"It sure is, kid… glad to see you're both okay." Dan and Mary both ran towards him and hugged him. After letting go of them, James turned to Jack, his face becoming much more serious.

"I think it's time you tell your story, Jack." He sighed and nodded in agreement.

"Right… it's certainly long overdue. Mary, there's something I need to tell you." Her face changed to one of curiosity as she walked over and sat down on the couch next to Jack. A moment later, Jack wrapped his arms around Mary and started to cry. "I'm so sorry… I should have stood up for you all those years ago. I'm sorry Mary." She turned to Dan and gave him a look that said *what's with this guy?*

"Are—are you okay, Jack?" asked Mary, finally deciding to reciprocate the hug. Jack let go of Mary and wiped the tears from his eyes.

"Forgive me, this isn't easy. Please understand, what I'm about to tell you is something I deeply regret, and I want you to know that I'm very sorry for what happened. Mary…" He took her hand in his as he tried to maintain his composure enough to continue talking. "My name is Jack Michael Stevenson, and I'm your father." A visible wave of relief could be seen on his face after finally telling her the truth. "I imagine you have quite a few questions… I'll answer them if I can." Mary sat on the couch, unsure of how to feel about what was just said. Dan knew this part but was still worried that Mary wouldn't respond well to the news.

"Y—You're my father? How? I don't understand, you're way older than my dad from Redfield. You can't be him." Jack nodded.

"I know this is confusing, trust me, but it's true. The me you've spent your life with back in 1965 is merely a representation of me. I designed that computer simulation." Jack realized he would need to tell her everything. "Okay, so maybe I need to explain more."

With that said, Jack proceeded to tell both Dan and Mary the story he told James when they came back from Level 0. It wasn't easy, but he told them everything, making sure to leave nothing out. Once he was done, a silence fell over the office for at least a minute while they tried to process everything that was said.

"Jesus," said Dan, almost in disbelief of Jack's story. "I always had gotten a bad feeling from Robert and Sebastian, but nothing on this level. I'm sorry you had to go through that." Jack's face revealed a hint of a smile.

"Thank you, Dan, I appreciate that. Mary, are you okay sweetie?" He put his hand lightly on her shoulder. She instantly leaned into him and wrapped her arms around him.

"I love you Dad." The words brought Jack to tears, and it brought a smile to both James and Dan who were looking on at this very special moment. "It was hard to hear what you had to say, but I understand. I want you to know there's no hard feelings from me. It must have been so hard being in that position—I'm sorry Dad. Thank you for being honest with me." Jack fell into Mary's arms and shuddered with the release of decades worth of guilt.

"Oh Mary, you have no idea how much that means to me.

I've spent the last thirty years regretting my decision and wishing with all that I am that I could take it back. Thank you for understanding, and for giving an old man hope. I love you." He looked over at Dan, who was admiring the emotionally packed scene taking place. "Thank you for always being there for my girl, Dan. James has told me about your relationship with her—you're a fine young man and I'd be proud to one day call you my son-in-law."

Mary got up from the couch and returned back to Dan's side, giving him a kiss and taking hold of his hand. Jack smiled at this and then turned his attention to James.

"And you, James... well, I don't think I can possibly thank you enough. You've saved these two so many times, even in the bleakest of situations. You have my eternal gratitude." As always, James remained modest in his response.

"I was just doing what was right, Jack. I couldn't let Dan and Mary die right in front of me—I just couldn't."

"Like I said earlier, you're a better man than I ever was. Don't let that ever change."

4

"Hang on," said Dan. "How did we manage to survive? If I understood it correctly, that Myriad thing was designed to wipe everything—including us." There was a brief silence in the office before Dan turned towards James. "You saved us again, didn't you? You found a way to bypass it, right?"

"I didn't bypass it, exactly... the other 2,000 kids were still wiped from existence. I was only able to protect you two from the code—I just didn't have enough time to save the others."

"How did you manage to do that?" asked Jack. "I don't see how you could write something so complex, so quickly."

James took another sip of his coffee and took a moment to collect his thoughts.

"So, I'll admit, there was a few minutes where I was out of hope… I had no ideas left and I thought I'd have to watch the two of you die in front of me. It was actually you, Jack, who had put the idea into my head. At the end of your explanation of the Myriad Code, you mentioned that it didn't touch the elevator system so that there was still a way out.

"I knew it was a long shot, especially given the time constraint, but I had to try. I quickly logged into both of the primary servers that were running their protocols and initiated a transfer to the elevator's computer system. Literally seconds before you executed the code, the transfer finished, so I had no time to warn either of you that I would be shutting down your protocols—sorry about that.

"I was hoping I didn't have to shut them down at all, but unfortunately while they're active, a couple of files have locks on them which was preventing me from copying them. Once I shutdown your protocols I ran out of time, so I'm hoping the two files I missed weren't vital. I'm guessing they were memories from a few minutes prior to the clean slate."

"My god, James," said Jack. "Seriously, your brain is scary sometimes." James chuckled and shook his head.

"You were the one who gave me the idea… even if you didn't intend to."

"You've always been modest," said Dan. "Seriously, you're the smartest guy I've ever met. Thank you again for saving us."

James turned back to face Dan and Mary, who were still locked hand in hand.

"So, there's one last thing. We need to figure out what you two want to do as far as your future." An odd, confused look came over Dan's face.

"What do you mean? Can't we just live together like we have been?" *I should have known. You're an amazing person, Dan. You're going to make a great husband and father someday.*

"Yes, of course you two can stay together. I wouldn't dream of separating you… you're perfect for one another. I just meant, do you want to return to the simulation of 1965 and live out your life there, or would you prefer us to design a more modern one that's based off of the current year." The expression on both of their faces made it clear that neither of them had thought of that.

"Regardless of which we choose, will we still be able to come visit you two?" asked Mary. James laughed.

"Absolutely. I plan to copy whichever simulation you choose to the private server I have and will make sure to keep an eye on things as well, just in case. I also plan to make a watch for both of you, similar to the one I originally gave Dan, that would give you the ability to return to the real world whenever you'd like." Their faces were overjoyed.

"That'd be awesome!" said Dan. "Let's do it. Oh, wait… we still have to decide which simulation we want. Mary, what do you think? I know you have your family in 1965, but I'm sure between James and Jack, they could create an updated one as well." Both of them nodded, agreeing that it could be done.

"Hmm, I think we should go with a new one. You've spent so much time with me in my time, it's only fair that I spend time with you in yours." Dan smiled and kissed her.

"Updated simulation it is," said James. Jack and I will start working on this right away. Keep in mind, it'll probably take a while as I want to make sure it's capable of accommodating your life for the foreseeable future."

For the next two months, James and Jack worked diligently on creating an updated simulation for Dan and Mary to live out their lives together. They completed the new simulation on February 10th, 2005—thirty-six years to the day since Mary had become patient zero.

6

Mary and Dan had spent the last year enjoying their new lives together in the updated simulation of Redfield. Due to the large time lapse between Mary and Dan's worlds, the simulation was developed in a way that omitted Mary's parents—with her approval, of course. Both of Dan's parents were included in the simulation, and they were modeled after his parents from the current year. With the watches that James had designed, Mary was able to come visit her real-life dad whenever she wanted.

Before the two of them were put into the simulation for the first time, James had gone over all the added functionality to the watches, which was quite extensive. The two of them more or less had complete control of the simulation from within. James and Jack had taken no shortcuts in redesigning the simulation, so the realism had been significantly improved in comparison to the one in 1965.

Both of Dan's parents instantly fell in love with Mary, which was a mutual feeling for her with them. Parts of the simulation had been reused, so they still lived in the large brick house on the corner, across from the school. The school, however, had been completely redesigned to better serve the students.

Jack Stevenson returned to his solitary life in Redfield. He took a part-time job at the local school, helping out in the computer department. Money wasn't an issue, but it helped pass the time and he enjoyed helping the kids with their computer classes. He invited James to come and show the students some of the work he had done, which was a great hit with them.

It took some convincing, but Jack eventually managed to convince his ex-wife, Helen, to arrange a meeting with Mary and Dan—James was invited too. The five of them met at Jack's house and for the most part, the evening went well. There was a fair amount of explanation required, but Helen took it much better than anyone thought she would.

After the events of Project Destiny, James decided to mostly retire from the computer scene. He still dabbled in programming from time to time, working on small projects at home, but as for finding another job, he didn't see it necessary. With the salary he had made during the project, he could live on the savings he had built up over the years.

Even two years later, he still found himself wondering what became of the AI. He felt she had been caught off-guard by the Myriad Code, but at the same time, he couldn't shake the feeling that she was still around—somewhere. Regardless, he always keeps a protocol running on his private server to monitor the activity from The Elysiam.

Mary and Dan, now both eighteen, had decided to get married. The wedding took place on October 31st, 2006. They had reached out to James to see if it was possible for Jack and himself to attend, which of course he had found a way to make it work. The wedding was a beautifully spooky,

Halloween themed day with Mary and Dan surrounded by family and a few friends.

The kind of love they shared was a very rare and special thing. Through even the darkest and weirdest of times, their love never faltered. Even while at the mercy of a self-aware AI, the two of them found a way back to one another. Mary and Dan were living their happily ever after, and thanks to James, the four of them would never be apart.

Time was indeed funny sometimes—but this time, at least, it seemed very right.

EPILOGUE

PROJECT DESTINY WAS OVER. IT HAD BEEN OFFICIALLY SHUT down, and all evidence, both digital and physical had been destroyed. The execution of the Myriad Code had taken care of all the existing protocols, which stopped the AI from revealing the long-hidden secrets of The Elysiam to the public. As for the AI, there was never any research done to determine her fate. It's unlikely she was able to escape the clean slate protocol, but James knew better than to ever doubt the intelligence and power of her again.

In the depths of Level 0, an eerie silence continued to dominate the area, which had previously been filled with the whirring sound of servers and medical equipment. The highly secured elevator served as the single entry and exit point, so when the Myriad Code was executed, the elevator remained online and unaffected. In the quiet darkness of the office, the floor indicator above the elevator blinked to life and displayed a single image.

☺

NOTES

CHAPTER I

1. RSA is a cryptosystem used to secure data transmission by creating a public key based off of two large prime numbers. These are kept secret as they're the only reliable way to decrypt the data.

Made in the USA
Monee, IL
08 November 2026

67379969R00121